Senescence

Howard Thomas

Book title: Senescence
Author: Howard Thomas, www.sidthomas.net/wp

Cover design and illustrations: Debbie Maizels, www.zoobotanica.com
Published by Howard Thomas, www.plantsenescence.org
Printer: Cambrian Printers, Aberystwyth
Publication date: 2016 (first edition)

For Helen and Ben, whether they want it or not

iv

Bookmap

Chapter 1 In the beginning

SEN   SEN

SE'NATE. *n. f.* [*senatus*, Latin; *senat*, French.] An assembly of counsellors; a body of men set apart to consult for the publick good.

> We debate
> The nature of our seats, which will in time break ope
> The locks o' th' *senate*, and bring in the crows
> To peck the eagles.   *Shak. Coriolanus.*
>   There they shall found
> Their government, and their great *senate* chuse. *Milton.*
> He had not us'd excurions, spears, or darts,
> But counsel, order, and such aged arts ;
> Which, if our ancestors had not retain'd,
> The *senate's* name our council had not gain'd. *Denham.*
> Gallus was welcom'd to the sacred strand,
> The *senate* rising to salute their guest. *Dryden.*

SE'NATEHOUSE. *n. f.* [*senate* and *house*] Place of publick council.

> The nobles in great earnestness are going
> All to the *senatehouse*; some news is come. *Shakespeare.*

SE'NATOR. *n. f.* [*senator*, Latin; *senateur*, French.] A publick counsellor.

> Most unwise patricians,
> You grave but reckless *senators.* *Shakesp. Coriolanus.*
> As if to ev'ry fop it might belong,
> Like *senators*, to censure, right or wrong. *Granville.*

SENATO'RIAL. } *adj.* [*senatorius*, Lat. *senatorial*, *senatorien*, Fr.]
SENATO'RIAN. } Belonging to senators ; befitting senators.

To SEND. *v. a.* [*sandgan*, Gothick ; *þenƿan*, Saxon ; *senden*, Dutch.]

1. To dispatch from one place to another.
> There shalt thou serve thine enemies, which the Lord shall *send* against thee, in hunger and in thirst. *Deutr.* xxviii. 48.
> *Send* our brother with us, and we will go down. *Gen.* xliii.
> His citizens *sent* a message after him, saying, we will not have this man to reign over us. *Lu.* xix. 14.
> The messenger came, and shewed David all that Joab had *sent* him for.   2 *Sa.* xi. 22.
> My overshadowing spirit and might with thee
> I *send* along. *Milton.*
> His wounded men he first *sends* off to shore. *Dryden.*
> Servants, *sent* on messages, stay out somewhat longer than the message requires. *Swift.*

2. To commission by authority to go and act.
> There have been commissions
> *Sent* down among them, which have flow'd the heart
> Of all their loyalties. *Shakesp. Henry VIII.*

3. To grant as from a distant place : as, if God *send* life.
> I pray thee *send* me good speed this day, and shew kindness unto my master. *Gen.* xxiv. 12.
> O *send* out thy light and thy truth ; let them lead me. *Ps.*

4. To inflict ; as from a distance
> The Lord shall *send* upon thee cursing, vexation, and rebuke, in all that thou settest thine hand unto. *Deutr.* xxviii.

5. To emit ; to immit ; to produce.
> The water *sends* forth plants that have no roots fixed in the bottom, being almost but leaves. *Bacon's Nat. History.*
> The senses *send* in only the influxes of material things, and the imagination and memory present only their pictures or images, when the objects themselves are absent. *Cheyne.*

6. To diffuse ; to propagate.
> When the fury took her stand on high,
> A hiss from all the snaky tire went round :
> The dreadful signal all the rocks rebound, }
> And through the Achaian cities *send* the found. *Pope.*

7. To let fly ; to cast or shoot.

To SEND. *v. n.*

1. To deliver or dispatch a message.
> I have made bold to *send* in to your wife :
> My suit is that the will to Desdemona
> Procure me some access. *Shakesp. Othello.*
> They could not attempt their perfect reformation in church and state, 'till those votes were utterly abolished ; therefore they *sent* the same day again to the king. *Clarendon.*

2. *To Send for.* To require by message to come, or cause to be brought.
> Go with me some few of you, and see the place ; and then you may *send for* your sick, which bring on land. *Bacon.*
> He *sent for* me ; and, while I rain'd my head,
> He threw his aged arms about my neck,
> And, seeing that I wept, he press'd me close. *Dryden.*

SE'NDER. *n. f.* [from *send*.] He that sends.
> This was a merry message.
> —We hope to make the *sender* blush at it. *Shak. H. V.*
> Love that comes too late,
> Like a remorseful pardon slowly carried,
> To the great *sender* turns a sour offence. *Shakespeare.*
> Rest with the best, the *sender*, not the sent. *Milton.*

SENE'SCENCE. *n. f.* [*senesco*, Latin.] The state of growing old ; decay by time.
> The earth and all things will continue in the state wherein they now are, without the least *senescence* or decay, without jarring, disorder, or invasion of one another. *Woodward.*

SE'NESCHAL. *n. f.* [*seneschal*, French, of uncertain original.]

1. One who had in great houses the care of feasts, or domestick ceremonies.
> John earl of Huntingdon, under his seal of arms, made sir John Arundel, of Tretice, *seneschal* of his houshold, as well in peace as in war. *Carew's Survey of Cornwall.*
>   Marshal'd feast,
> Serv'd up in hall with sewers and *seneschals* ;
> The skill of artifice, or office, mean! *Milton's Par. Lost.*
>   The *seneschal* rebuk'd, in haste withdrew ;
> With equal haste a menial train ; ensue. *Pope's Odyssey.*

2. It afterwards came to signify other offices.

SE'NGREEN. *n. f.* A plant. *Ainsworth.*

SE'NILE. *adj.* [*senilis*, Latin.] Belonging to old age ; consequent on old age.
> My green youth made me very unripe for a task of that nature, whose difficulty requires that it should be handled by a person in whom nature, education, and time have happily matched a *senile* maturity of judgment with youthful vigour of fancy. *Boyle on Colours.*

SE'NIOR. *n. f.* [*senior*, Latin.]

1. One older than another ; one who on account of longer time has some superiority.
> How can you admit your *seniors* to the examination or allowing of them, not only being inferior in office and calling, but in gifts also ? *Whitgifte.*

2. An aged person.
> A senior of the place replies,
> Well read, and curious of antiquities. *Dryden.*

SENIO'RITY. *n. f.* [from *senior*.] Eldership ; priority of birth.
> As in all civil insurrections the ringleader is looked on with a peculiar severity, so, in this case, the first provoker has, by his *seniority* and primogeniture, a double portion of the guilt. *Government of the Tongue.*
> He was the elder brother, and Ulysses might be consigned to his care, by the right due to his *seniority.* *Broome.*

SE'NNA. *n. f.* [*sena*, Latin.] A physical tree.
> The flower, for the most part, consists of five leaves, which are placed orbicularly, and expand in form of a rose : the pointal afterwards becomes a plain, incurved, bivalve pod, which is full of seeds, each being separated by a double thin membrane. The species are three. The third sort, that used in medicine, is at present very rare. *Miller.*
> What rhubarb, *senna*, or what purgative drug,
> Would scour these English hence? *Shak. Macbeth.*
> *Senna* tree is of two sorts : the bastard *senna*, and the scorpion *senna*, both which yield a pleasant leaf and flower. *Mort.*

SE'NNIGHT. *n. f.* [Contracted from *sevennight.*] The space of seven nights and days ; a week. See FORTNIGHT.
> Time trots hard with a young maid between the contract of her marriage and the day it is solemnized : if the interim be but a *sennight*, time's pace is so hard that it seems the length of seven years. *Shakesp. As you like it.*

SENO'CULAR. *adj.* [*seni* and *oculus*, Latin.] Having six eyes.
> Most animals are binocular, spiders octonocular, and some *senocular.* *Derham's Physico-Theology.*

SENSA'TION. *n. f.* [*sensation*, French ; *sensatus*, school Latin.] Perception by means of the senses.
> Diversity of constitution, or other circumstances, vary the *sensations* ; and to them of Java pepper is cold. *Glanv. Sceps.*
> The brain, distempered by a cold, bearing against the root of the auditory nerve, and protracted to the tympanum, causes the *sensation* of noise. *Harvey on Consumptions.*
> This great source of most of the ideas we have, depending wholly upon our senses, and derived by them to the understanding, I call *sensation.* *Locke.*
> When we are asleep, joy and sorrow give us more vigorous *sensations* of pain or pleasure than at any other time. *Addison.*
> The happiest, upon a fair estimate, have stronger *sensations* of pain than pleasure. *Rogers.*

SENSE. *n. f.* [*sens*, French ; *sensus*, Latin.]

1. Faculty or power by which external objects are perceived ; the sight ; touch ; hearing ; smell ; taste.
> This pow'r is *sense*, which from abroad doth bring
> The colour, taste, and touch, and scent, and found,
> The quantity and shape of ev'ry thing
> Within earth's centre, or heav'n's circle found :
> And though things sensible be numberless,
> But only five the *sense's* organs be ;
> And in those five, all things their forms express,
> Which we ran touch, taste, feel, or hear or see. *Davies.*
> Then is the foul a nature, which contains
> The pow'r of *sense* within a greater pow'r,
> Which doth employ and use the *sense's* pains ;
> But fits and rules within her private pow'r. *Davies.*
>   Both contain
> Within them ev'ry lower faculty
> Of *sense*, whereby they hear, see, smell, touch, taste. *Milt.*
> Of the five *senses*, two are usually and most properly called the *senses* of learning, as being most capable of receiving communication of thought and notions by selected signs ; and these are hearing and seeing. *Holder's Elements of Speech.*
>   There's

A facsimile of the entry for 'Senescence' in Samuel Johnson's *A Dictionary of the English Language* (1755), p. 1787[1]

JOHNSON ...Make a large book; a folio.
BOSWELL But of what use will it be, Sir?
JOHNSON Never mind the use; do it.

Does the world need another book? Dr Johnson[2] thought it did, and who would argue with the Great Man ('Love of Dr Johnson' wrote Dashiell Hammett[3], 'is the mark of the pathologically meek'). But a book about senescence? Well, it's true that there is a vast literature about death and the events leading to it. Ageing is nothing more than the biological response to the passage of time. You could say, therefore, that virtually the entire body of literature addresses this subject in one way or another. Like Art, Life in all its forms tells a bigger story about time and change than can any single human's experience. It's this belief that has made me (meekly) heed Dr Johnson's call.

Senescence is part of a cloud, or perhaps fog, of terms referring generally to the process or condition of growing old.[4] A *Thesaurus* search for 'senescence' reveals words for maturity, ripeness, seniority and longevity, but the dominant associations are with notions of decay, decline, gerontology, morbidity and mortality. This reflects the etymological origin of the word (from Latin *senescere* to grow old) and its association with senility and the medical problems of human ageing. The earliest use of 'senescence' in print, cited by Samuel Johnson and recorded by the *Oxford English Dictionary* (OED 2010), was in 1695, in John Woodward's *An Essay toward a Natural History of the Earth and Terrestrial Bodies, especially minerals, &c.* Woodward reflected that it was reasonable to conclude that a divine power, having provided a planet perfectly suited to the needs of its human inhabitants, would 'continue to preserve this Earth, to be a convenient Habitation for the future Races of Mankind' - it followed that 'the Earth, Sea, and all natural things will continue in the state wherein they now are, without the least Senescence or Decay'. Here, as is so often the case, senescence is coupled with a term for deterioration, dissolution or loss of potency, acquiring a similar meaning by association.

The first uses of 'senescence' in a genuinely biological context occurred during the 1870s and were almost exclusively confined to the animal kingdom. Charles Sedgwick Minot was a particularly enthusiastic adopter of the term, and used it in the title of his 1891 article *Senescence and Rejuvenation* (Journal of Physiology 12: 97-192). This may have been the first in an intended series under that title which was never completed, since it consists largely of careful and detailed measurements on weight gain in guinea-pigs. In his 1908 book *The Problem of Age, Growth, and Death. A Study of Cytomorphosis*, Minot set forth his definition of the word: 'With each successive generation of cells the power of growth diminishes... This loss of power I term senescence'.

In retrospect, it is perhaps unfortunate that botany has chosen to adopt the term 'senescence' to describe a phase of plant development which, while indeed characteristic of older leaves and other organs, has its own unique properties and does not represent inevitable and irreversible decline into elderly decrepitude, impotence and death. That 'senescent' does not necessarily signify a moribund or decrepit state was asserted by Bishop William Stubbs (1886) in *Seventeen lectures on the study of mediaeval and modern history* (OED 2010): 'It is not a dead but a living language, senescent, perhaps, but in a green old age'. Current physiological understanding of the senescence condition and its positive roles in plant growth, differentiation, adaptation, survival and reproduction, follows Stubbs in acknowledging senescence to be a phase of development that follows the completion of growth, is absolutely dependent on cell viability and which may or may not be succeeded by death.

Unfriendly friendly universe
Edwin Muir[5]

Someone once said that all poetry is about birds. Or, to generalise further, at the heart of poetry is the revelatory lift that comes from being shown, through metaphor, simile and allegory, that something is like something else (often a bird, for sure, but an autumn leaf or a fading petal will do). It's a harmless conceit to project death, life, and whatever comes between onto the fate of the entire universe and everything in it, so let's do just that. At the birth of the universe, space-time was almost infinitesimally small. So small that there was only room for photons of almost infinitesimally short wavelengths. Wavelength and energy are inversely related. Accordingly, energies in the new-born universe were almost infinitely large. As the universe expanded, wavelengths could relax, and energy levels came down. Today the peak wavelength of cosmic background radiation is a little over 1 millimetre. This corresponds to an energy of about 2.7°C above absolute zero. The cosmos is old and cold, and getting older and colder all the time.

Since the Big Bang, photons have been on a one-way journey through the spectrum, from short wavelength and energetic to long wavelength and frigid.[6] But the Universe is patchy. Galaxies and other objects are lumps of mass-energy caught in eddies where the flow has locally slowed down. In our corner of our galaxy, we experience a small window in the full photon energy range. The band of radiation from our star, the sun, is further filtered by our planet's atmosphere before it arrives at the Earth's surface, and by our eyes before it reaches our brains. When I was very young, a lighthouse keeper in Devon taught me the colours of the rainbow. 'Run Off You Girls, Boys In View – them are all the colours we can see' he said, and I never forgot it. The wavelengths we perceive lie between 380 nanometres (violet) and 760 nanometres (far red). A nanometre is one thousand-millionth of a metre. When a photon within the range of the visible spectrum interacts with matter, it can make an electron jump to a higher energy level before eventually coming down again. All life is made possible by electrons within atoms moving in this way, from one energy level to another.

Fall, little leaf,
Pit-pit-a-pat,
And settle on
My father's hat.

Unlike JB Morton (Beachcomber)[7], I do not recall my father ever wearing a hat. He kept a vigorous head of hair until the end of his life and seems never to have felt the need to cover up. Thus far I appear to take after him in this regard. I also followed him by going grey quite early, beginning in my fourth decade. Fading hair, like the fading colour of Morton's autumn leaf, is said to be a melancholy reminder of the passage of time. When Samuel Beckett, referring to the colours of Watt's hat and coat, writes of 'time, that lightens what is dark, and darkens what is light', he was expressing a universal, literal truth. Leaves and hair are microcosms that, in a curious way, recapitulate the universe's slide down the gradient of photon energy.

We can learn from plants.[8] By being at once unlike and like us, they can teach us what in senescence is common to life in general, and what is merely the preoccupation of our own self-absorbed species. The title says that this book is about 'senescence', and not 'plant senescence'. I make no apology about omitting the qualifier, for the following reasons. First, plants are where I start from, because I happen to be a plant biologist, but this book roams freely across the whole of biology and beyond. The second reason follows from the first: looking out on the great expanse of biological ageing from the standpoint of the green world I'm familiar with, it seems to me that canonical understanding of what happens in the period between life and death is conceptually narrow and incomplete. Third, if you opened this book in the expectation of an account of human ageing, you may be disappointed, but you shouldn't be. 'Even the wisest among you is only a conflict and hybrid of plant and ghost.' Thus Spake Zarathustra. And anyway, it does all of us good to consume our greens.[9]

Chapter 2 Growth and change

Turnover[10]

It is worth dying to find out what life is.
    TS Eliot[11]

Life has a beginning, a middle and an end. Sometimes the middle goes missing. The end may come unbidden at any time. It might be traumatic or pathological – the consequence of a chance occurrence. Or it could be by design. The variety of living things is reflected in the many different modes of death awaiting them. We gain a deeper understanding of life by observing how and why it comes to a conclusion.

All living organisms must practice self-preservation. As Michael Crichton wrote in *Congo*: 'The purpose of life is to stay alive'. Some kind of perception of death as a state to be avoided seems to be universal. But as far as we know, humans are the only creatures with an existential awareness of their own mortality. Ageing, senility, failing powers, declining vitality and ultimate oblivion are human fixations. Like much else about the human condition, they have become medicalised. We are told to believe that ageing is a curable disease. The preoccupations and insights of gerontology, the biomedical study of old age in humans, have come to dominate our conceptual grasp of the mechanisms and meanings of life's end in the natural world.

It's a pity, but understandable, that Biology has become redefined as Medical Science. The Human Genome Project ignited a culture change across the life sciences. Biology is dominated and redefined by the sheer volume of medical research. Huge sums of money gush from governments, charities, insurance companies and the biotechnology industries, in the name of health care. But it's worth asking whether an obsession with the science of ill-health is healthy. We may connect this fixation with what Robert Pogue Harrison, in his book *Juvenescence: A Cultural History of Our Age,*[12] refers to as 'the storm of juvenescence that has swept us up in the past several decades' and question, with Harrison, whether the outcome will be 'a genuine rejuvenation or a mere juvenilization of culture'. Bertrand Russell, in a letter to Warren Allen Smith in 1956, wrote that he regarded human beings to be 'a trivial accident, which would be regrettable if it were not so unimportant'. Just as giving healthcare pride of place in politics is the signature of a society defining itself as sick, so too placing the human experience at the conceptual heart of contemporary biology is bad science. A human is a complicated organism, perhaps the most complicated in the natural world. Generalising from complexity is a dangerous strategy. Physicists built their conceptual model of the structure of matter by starting with the hydrogen atom, not californium.

Equating biology with medicine breeds parochialism. Principles of genetics and evolution are subject to narcissistic subversion. Genes determine how 'we' develop, what diseases 'we' will get and what medicines 'we' will need. Health is the absence of ill-health. Life has become redefined as the fight (ultimately lost) with pathology and death. This feels like reversion to the proto-medical concept of the four humours, specifically the melancholic temperament and its preoccupation with winter and old age. Quite often, 'new' insights from biomedical study are not new at all to those familiar with the comparatively unregarded hinterlands of biology. In fact, sometimes generalizations made on the basis of such insights are just plain wrong. Of course, gerontological ideas about the endgame must be acknowledged. But most biology is not human biology. The scientific landscape may be dominated by the dense forest of human ageing and healthcare, but there is a much wider and richer natural panorama in the world beyond. 'Nature is the music, human is the static', as John Updike almost said.

If we manage to avoid fatal disease or accident, we will grow old. Then we will die because we got old. In this context, the term ageing is synonymous with growing old and the human experience of deterioration leading to death. Of course, in the long run, errors will build up and living cells and tissues will show signs of wear and tear. In the general biological sense, however, ageing is more than this. The word is usefully applied to all changes that occur with time. It therefore embraces the time-based processes of growth and differentiation as well as maturity and mortality. Accordingly, ageing is not simply another name for declining viability. The notion that we start to die as soon as we are born is not helpful in understanding the biology of ageing. In the words of Robert Pogue Harrison, 'nothing in the universe – be it the newborn infant or the universe itself – is without age. If a phenomenon does not age it is not of this world; and if it is not of this world, it is not a phenomenon'.[13]

Entropy is disorder. 'Things fall apart; the centre cannot hold'.[14] In a famous passage from *What is life?*, Erwin Schrödinger answers the question posed by the title of his influential little book thus: 'a living organism continually increases its entropy - or, as you may say, produces positive entropy - and thus tends to approach the dangerous state of maximum entropy, which is of death. It can only keep aloof from it, i.e. alive, by continually drawing from its environment negative entropy'.[15] A biologist recognises 'negative entropy' as *development*, the general term for the changes in form and function brought about through growth and differentiation. To develop takes work. As Anton Chekhov observed, 'only entropy comes easy'.

We do not grow absolutely, chronologically. We grow sometimes in one dimension, and not in another; unevenly. We grow partially. We are relative. We are mature in one realm, childish in another. The past, present, and future mingle and pull us backward, forward, or fix us in the present. We are made up of layers, cells, constellations.

Anaïs Nin *Diaries* Volume 4 (1944-1947)[16]

Populations grow. We may think of an individual as a population of one. In a typical growing biological system, the rate of increase at any particular moment is directly related to the size of the population. This sort of relationship is said to be density-dependent. The classical S-shaped (sigmoidal) growth curve emerges from the mathematics of density dependence.[17] Growth begins slowly when population size is small. Its rate is greatest when density is optimal with respect to physiological and environmental constraints. Subsequently growth wanes as limiting external and internal factors become increasingly influential. Finally the plateau of maximal size is approached. Growth is the consequence of one or both of two activities. Cells increase in number, or they increase in size. Plant cells are born in meristems, centres of mitosis (cell division) at the tips of shoots and roots and other locations. Cell proliferation is usually an early event in plant growth and development. Subsequently cells expand by the influx of water. Post-mitotic growth is hydraulically powered. Plant form at maturity is sustained by water in cells exerting roughly the same pressure per unit area as does the air in a car tyre.

Growth curves vary from tissue to tissue in their proportions and their relations to time. Senescence responses are correspondingly diverse. The total quantity and annual biological production of wood in the biosphere are estimated to be a trillion (10^{12}) tons and 10 billion (10^{10}) tons respectively. This is senescence and death on a huge scale: wood largely consists of dead cells that achieved full size and morphological maturity comparatively quickly after they were formed. Senescence followed by death of cell contents were completed soon after growth levelled off.[18] By contrast, the cells bordering the pores (stomata) that allow passage of gases and water vapour into and out of leaves have a different schedule.[19] A recognizable senescence program may not be initiated until long after stomatal cells reach structural and functional maturity. In some species, stomatal cells remain in the pre-senescent state when the leaf as a whole is shed. Death makes the form of the tree in winter.

Differentiation is the change in structure and function that results in cell, tissue and organ specialization. Senescence and differentiation are intimately connected; in the words of Caleb E Finch 'the extent of intrinsicality in cell senescence should be viewed as the outcome of cell differentiation'[20]; and, we might add, vice versa. Senescence is a normal and even essential feature of the post-mitotic phase of the cell life-cycle. It is immediately preceded by (and sometimes partially overlaps with) the growth period, and often specifies cell fate. Senescence is triggered differentially in tissues and organs, resulting in complex anatomies and morphologies that change and adapt over time. It is the means by which resources are recycled from obsolete body parts to new developing structures. And variations on the theme of programmed senescence have been shaped by evolution to give rise to a diversity of structures within the lifecycle.[21] Variety may be the spice of life, but there can be no variety without death.

If you have lower than a ten percent turnover, there is a problem. And if you have higher than, say 20%, there is a problem.
Richard M. Nixon[22]

Increase, maturity and decline are scalable properties of populations. These elements of population dynamics are broadly conserved across taxonomic kingdoms and levels of organization. Density-dependence, interacting with demographic structure, has an important influence on the evolution of optimal life histories, lifespan and senescence. And recent research in a range of organisms is beginning to suggest that the factor linking growth, development and senescence is nutrition. Molecules, cells, tissues, individuals, and even whole biomes, turn over as they progress through the phases of growth and decline. Turnover, which is defined as flux through a pool, is more or less synonymous with ageing. In biology, the pool may be a population of molecules, cells, tissues, organs, organisms and may be extended to include whole faunas, floras and biomes. Flux means rate of flow; in this case, flow into and out of the pool. Members of the population are recruited, remain in the pool for some period and then exit. The theory of turnover can be mathematically quite challenging. But the principles can be understood without recourse to differential equations and the like. The important factors in turnover are the size of the pool and the rates at which individuals enter and leave it. If entry rates exceed exit, the pool grows. If entry and exit rates are balanced, pool size remains the same and the system is said to be at the steady state.

Karl Marx[23] defined turnover as 'a circuit performed by a capital and meant to be a periodical process, not an individual act.' He thought of turnover as dynamic: 'the duration of this turnover is determined by the sum of its time of production and its time of circulation'. The relationship between turnover and ageing is realised in the structural dynamics of the plant lifecycle.[24] Turning over a new leaf is a way of life for plants. The canopy – that is, the mass of foliage on the shoot - represents a pool of organs: Marx's 'capital' in the economy of the plant. Newly initiated leaves join the pool, grow and mature, become senescent and ultimately die, and leave the pool. When the plant is young, more leaves are initiated than become senescent. The total number of leaves on the shoot – the pool – increases. As the plant reaches maturity, leaf initiation declines and the pool of foliage enters the steady state. Later, foliage is lost, without replacement, through senescence and death. The result is a net decrease in leaf population and eventually, as we see in annual plants or deciduous trees at the end of the season, the pool size becomes zero.

Chapter 3 Structures and stresses

Plant structure is modular[25]

Biological structure and function are organised hierarchically. Ageing and turnover reflect this. There are different orders of ageing going on in the life of a plant. One concerns the change over time in the state of the individual component of the pool of foliage. In this case, each leaf goes through an ageing sequence, starting as a newly-emerged and growing organ, progressing to full-grown structural maturity and finally entering a terminal phase of senescence and death. And then there is the behaviour of the pool itself, from initiation to steady state to decline and ultimately to zero. The study of ageing is much concerned with how turnover scales between levels of biological organization, from individual to population. The issue of scaling poses one of the Big Questions of modern biology[26], and is one I return to repeatedly: is ageing of the whole organism an expression of decisive events happening in or to its constituent cells and tissues? Intuitively it would seem that it must be. But in practice it is extraordinarily difficult to make such a connection.

Weismann considered that death was programmed in multicellular beings through an intrinsic limitation in the ability of their somatic cells to multiply. That limitation was made possible by "the division of work" introduced by the first differentiation processes between somatic cells and germ cells. Only germ cells must imperatively be immortal.

André Klarsfeld and Frédéric Revah[27]

The open-ended nature of plant organization and development is very different from the way growing animals attain form and function, and has fundamental implications for senescence. A critical distinction, first elaborated by the evolutionary theorist August Weismann[28], concerns separation of germline (gamete-producing) from soma (non-reproductive) cells in the animal body. There is no such differentiation into germline and soma in plants. A plant cell that is not terminally differentiated is generally totipotent – it has the capacity to regenerate an entire plant by division and differentiation. The biomedical field of stem cell research[29] aims to imitate this trick in the much more recalcitrant cells of animals and humans. Weismann's ideas about ageing live on in the disposable soma theory proposed by the gerontologist Tom Kirkwood[30] and others. Plants represent a challenge to the generality of this concept as the underlying principle of biological ageing and senescence.

The plant body plan, unlike that of animals, develops by the fractal-like repetition[31] of homologous structural units (phytomers). A section of stem with a leaf and its associated basal bud is the phytomeric unit from which the plant shoot is constructed. The rich diversity of forms and life cycles arises by variations in how phytomers are arranged in space, or when they initiate, develop and senesce. This makes the individual plant seem like a colonial organism, a population of genetically equivalent structural modules simultaneously in resource competition and physiologically integrated with each other. Comparing patterns of ageing and senescence in plants with those of colonial animals reveals instructive similarities and differences.[32] Ageing communal polyps and sea squirts are able to reset the biological clock during the sexual phase of the lifecycle. Through the reductive mode of cell division (meiosis) followed by fertilization, sex (in animals and plants) is able to re-boot juvenility. But long-lived plants also seem to be able to do something similar in their meristems, perhaps because of the way mitosis is organised in these centres of somatic cell proliferation.

In its usual overbearing way, the health business has appropriated the notion of stress and made it a matter of self-absorption and the pathologising of everyday human life. But for biology in general, stress has a rich and deep meaning, rooted in physics and engineering. It describes an environmental influence that invokes a corresponding strain. Stress and strain can be difficult concepts. The engineer J E Gordon[33] described how one of his students found the whole business so distressing that she ran away and hid. Nevertheless, we must try to understand biological stress and strain if we are to appreciate how and why living things senesce. The environment is stressful when it deviates from the ideal. But, for a living organism, the environment is *always* non-optimal in one way or another. This may be because some material or energetic need is unmet or in excess, or because each of the various physiological systems that make up cells, organs, individuals and populations has its own optimum. In fact Life, sustained by Schrödinger's negative entropy[34], is absolutely dependent on non-optimality in the environment. An organism can only tell where (and perhaps even what) it is in space and time because its environment deviates from optimality. Kindness kills is a profound biological principle.

If people think nature is their friend, then they sure don't need an enemy.
Kurt Vonnegut[35]

The biologist needs to be careful not to think narrowly of stress in terms of *distress*, a kind of analogy with human psychological and physical reactions to the pressure of troubling circumstances, or the cries in the night of an infant protesting that its life isn't perfect. Biologists should approach stress and strain as engineers do. Optimal conditions are not stress-free. No less than for immobile individuals such as plants, stress and strain are a way of life even for creatures that can migrate, or at least run away, to avoid exposure to seasonal adversity. Whatever your lifestyle, the environment is a constant test of fitness. The inescapable cycle of day and night is a case in point: as organisms that 'feed on sunshine', plants must make relatively rapid physiological adjustments to an ever-changing light environment. In temperate regions, day length varies with the season and invokes longer-term changes in development. Not only do plants and animals react to deviations from optimal environmental conditions; they also use such fluctuations as a source of information to trigger adaptive changes in structure and function.

Thinking of relationships between living things and their environment in terms of the physical properties of materials under stress gives us a vocabulary for talking about the meaning of strain. Living systems are homeostatic – they resist change, tending to adjust to stress by minimising strain and maintaining equilibrium. A homeostatic system has three modes of response to disturbance. It may behave elastically, bouncing back and resuming its former state. Beyond the elasticity threshold it may be plastic, undergoing deformation before settling on a new stable configuration. When the limits of elastic and plastic resilience are exceeded, a catastrophic response results and the system becomes incoherent, entropy increases, and in the case of biology, death follows. Homeostatic adjustment of an individual organism in response to changing environmental factors is termed acclimation. An example of acclimation familiar to gardeners is the process known as hardening off, in which an otherwise vulnerable plant can be toughened up by exposing it to non-lethal levels of stress. To misquote Don Van Vliet (Captain Beefheart).[36] 'You strain yourself, you train yourself'. Constant or recurrent environmental challenge not only invokes acclimation; it may also exert selective pressure, driving the evolution of traits that increase fitness under stress. Such adjustments, which occur over many generations and across entire populations, are termed adaptations. Adaptive traits are genetically determined, constitutive characters, expressed whether the organism is stressed or not. A cactus is a spiny, leafless succulent whether it's growing in its natural arid habitat or on the windowsill of an air-conditioned office.

There's so much crap talked about tactics by people who barely know how to win at dominoes.
Brian Clough[37]

Acclimation acts over the short term in individuals. It represents a tactical reaction to stresses that are often unscheduled. In contrast, adaptation is strategic in nature and is the basis of population-scale adjustments. Acclimation responses generally involve physiological tuning, which may or may not require some genes to become active and others to switch off. Over the longer-term these adjustments can lead to significant changes in form and function. The traits acquired during acclimation are, in the vast majority of cases, reversible and not transmissible to progeny (though there is growing interest in epigenetic effects, where acclimation can be associated with permanent changes to the genome). By contrast, adaptation acts on genetic variation within a population, favouring those lineages best fitted to survive the selective stress, and results in irreversible changes to the genome.

How does senescence relate to the acclimation and adaptation modes of environmental interaction? Senescence is one of the tactics deployed when a random, unforeseeable stress is experienced. Seasonal or otherwise predictable environmental cues can trigger senescence as part of an adaptive strategy. When the speed and severity of a developing environmental stress outrun physiological capacity to resist or avoid harmful strain, the senescence phase of development is aborted and vital processes are diverted into the pathological pathways that lead rapidly to morbidity and death. Playing dead (the technical term is thanatosis[38]) is a common behavioural measure used by animals under threat (possums are famous for it). Plants do it too. A winter dormant tree looks like a dead tree.

Organisms deal with non-optimal environmental conditions in different ways. They may adopt a strategy of tolerance, building resilient structures and physiologies able to survive by withstanding stress. Or they may take the avoidance route, confining growth to relatively favourable periods in a fluctuating environment, through life cycle strategies that minimise exposure of vulnerable stages to extreme conditions. The capacity to survive different degrees of stress varies greatly between species, and among genotypes of a single species. Factors in the environment that influence growth, development and survival are broadly divided into biotic and abiotic. Biotic stresses arise directly or indirectly from interactions with other living organisms, including pathogens and predators, as well as humans through selection and pollution. Abiotic factors originate in experiences of physical, chemical and energetic conditions such as light, temperature, water and nutrients. Time is an abiotic stress of particular significance for ageing and senescence.

Chapter 4 Space, soils and seasons

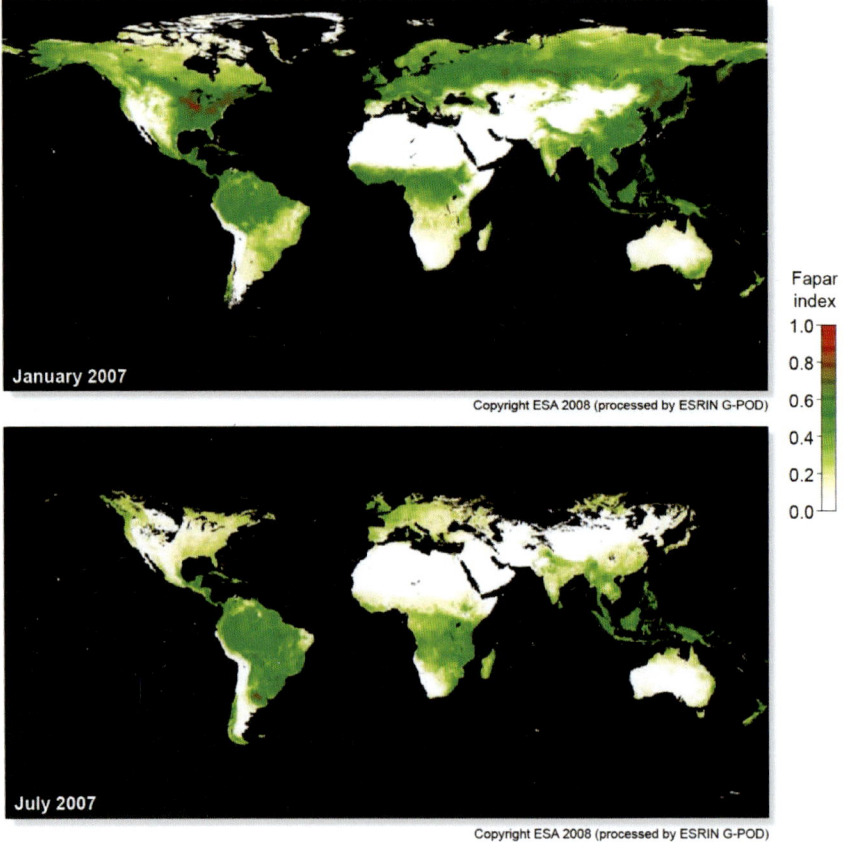

Monthly average global green biomass for January and July 2007, measured as Fapar (Fraction of Absorbed Photosynthetically Active Radiation), compiled from data collected by the Medium Resolution Imaging Spectrometer (MERIS) on board the European Environmental Satellite (ENVISAT)[39]

Astronauts describe Earth as the Blue Planet. For Earth Observation satellites, it's the Green Planet. Water and chlorophyll - the signatures of life in the solar system. Satellite images show the green colour of vegetation ebbing and flowing with the seasons. The annual cycle of greenness is the common experience of everyone living outside the tropics. Here the relationship between senescence and environment is simultaneously reactive and creative: a whole season of the year named for the fall of senescent foliage. The yellowing, withering and falling of leaves and other plant parts have been tropes for human ageing since poets and artists first engaged with what John Ruskin (1856)[40] called the 'Pathetic Fallacy'. In *Adonaïs*, his elegy on the death of John Keats (whose ode *To Autumn* is perhaps the most anthologised poem in the English language), Percy Bysshe Shelley wrote 'Grief made the young Spring wild, and she threw down/Her kindling buds, as if she Autumn were,/or they dead leaves'.[41] Autumnal imagery in the creative arts is echoed in the scientific term *apoptosis*, the Greek for falling off (of leaves), introduced in 1972 by John F R Kerr, Andrew H Wyllie and Alastair R Currie[42] to describe programmed cell death in humans and animals.

For eyes and cameras and spectrometers, registering the rhythmic changeability of vegetation is easy; but there is another world, largely hidden from sight: the rhizosphere. Senescence of root systems is at least as agriculturally and ecologically significant as senescence of above-ground parts. Root lifespan and the dynamics of subterranean tissue turnover have become a focus for research studies of carbon sequestration and climate change.[43] In the global carbon cycle, soil is one of the largest repositories (an estimated 2.5 trillion tons of carbon), and root turnover, determined by the lifespan of fast-cycling absorptive roots, is a major factor in the carbon, nutrient, and water cycles for plants and whole-ecosystems. Below-ground plant parts and their component cells and tissues grow, develop and senesce according to their own distinctive schedules, which usually do not relate in any simple fashion to those of the above-ground fraction. Roots go their own way. The roots of most species form beneficial associations (mycorrhiza) with soil fungi.[44] Truffles and boleti and death caps are the fruiting bodies of mycorrhizal fungi. Mycorrhizal associations tend to enhance root lifespan. On the other hand, environmental stresses such as drought or flooding promote root cell senescence. Root turnover is also sensitive to the availability of essential elements in the soil, particularly nitrogen and phosphorus. We will be hearing much more about the hidden world of root lifespan as climate change modelling becomes refined to incorporate concepts and data from intensifying research on the rhizosphere.

the beginning of autumn:
what is the fortune teller
looking so surprised at?
Buson[45]

At latitudes away from the equator, seasonal changes in daylength (photoperiod) and temperature are reliable environmental cues to prepare for the likely stresses to come. Many animals enter hibernation, aestivation, brumation or diapause to get them through the period of seasonal severity. Bumble bees hibernate, as noted by Truman Capote in Harold Arlen's charming song *A Sleeping Bee* ('When a bee lies sleeping/in the palm of your hand,/you're bewitched and deep in/love's long looked-after land').[46] The predictability of the seasons enables trees and other plants to make provision by deploying senescence strategically as a scheduled developmental event in the life-cycle. At the height of summer, trees show the first physiological inklings of the need to prepare for autumn and winter. This becomes a subtle example of pathetic fallacy in Johnny Mercer's English lyrics for *Autumn Leaves*, Joseph Kosma's classic *chanson*: 'Since you went away the days grow long/and soon I'll hear old winter's song;/but I miss you most of all my darling/when autumn leaves start to fall'.[47]

Just as hibernating animals do, plants build up reserves before entering dormancy. Proper timing of leaf senescence is vital if a plant is to balance its carbon and nitrogen demands as photosynthesis declines and nutrients are salvaged from yellowing foliage. Many species use daylength perception to initiate and sustain other developmental events, such as flowering. Sensitivity to the same environmental triggers allows major phases of the lifecycle to be strategically coordinated. The internal genetic and physiological pathways controlling such events are often, but not always, linked. The example of wheat - which together with rice and maize accounts for two-thirds of human food consumption – serves to show the agronomic value of closely associated traits under daylength control. The position in the wheat genome of a genetic locus that determines delayed leaf senescence is close to loci for photoperiod insensitivity and stature.[48] Studies of senescence in cereal species are beginning to reveal in molecular detail cross-talk between the regulatory pathways for photoperiodism, flowering, nutrient remobilization and grain fill. Modern highly productive, short-stemmed, non-lodging bread wheat varieties have been bred for extended greenness, dwarf habit and daylength insensitivity. The Green Revolution[49] of the 1940s to 1960s was made possible by such genetically linked traits moving together as a package during variety selection. You may amuse yourself by looking out for pre-Raphaelite actors drifting dreamily through fields of anachronistically dwarf cereals in film and advertising recreations of pre-20th century Arcadian nostalgia.

Trees! How ghastly!
Piet Mondrian[50]

On the campus of the University of Umeå in Sweden there's a specimen of European or quaking aspen (*Populus tremula*) that has become the most closely studied tree in the senescence business. Stefan Jansson and his colleagues have been analysing and documenting the seasonal behaviour of this single tree since 1999. This is heroic work. In an impatient scientific world, trees are not generally favoured as experimental subjects. Growth and development proceed at too leisurely a pace. Because mature trees don't easily fit into glasshouses or climate chambers, their environment can't be controlled and manipulated. A tree is too big and individualistic to cooperate with replicated experimentation. By contrast, plant science's lab rat, the cress-like weed Arabidopsis ('a crucifer with acolytes'[51]) lives fast (with a life-cycle seed to seed of a few weeks) and isn't too fussy about growing conditions. It's tiny - each individual takes up little more room than a champagne cork – and can be cultivated in large numbers in a small space. Just the experimental subject for an era in which research grants are limited in time and money.[52] Arabidopsis has contributed a lot to our understanding of senescence. But aspen has given new insights into what it means to be a long-lived species interacting with a complex and changeable environment.

All leaves of a mature aspen growing in boreal regions emerge from overwintering buds, grow to maturity and undergo autumnal senescence simultaneously. Development of the canopy (the entire mass of foliage) is synchronised, and follows a predictable timetable.[53] Regardless of weather and other environmental conditions, the Umeå tree initiates foliar senescence strictly according to photoperiod, on 10 September every year. To senesce on cue, the tree must already have developed the competence to respond to the daylength stimulus. It seems that this capability is acquired only after growth is arrested and winter buds are set, events that have been triggered by a light-sensitive pathway separate from that for senescence. Although the initiation of senescence can be predicted by date, the rate of the process once it is underway is influenced principally by temperature. The time from the onset of senescence until the leaves start to be visibly yellow could differ by up to 2 weeks, depending on how low the temperature falls during this period. Autumnal colour changes are enhanced by chilling and impeded by mild conditions. In years when the weeks immediately following senescence initiation are colder, the 'falling leaves of red and gold' (Mercer again) make an especially striking display.

Plants and light are locked together in a vital embrace. The green pigment chlorophyll collects the light energy that drives photosynthesis, the basis of all life on Earth. But plants have miniscule amounts of other pigments too, so-called photoreceptors that detect subtle variations in the light environment and send information to the physiological networks that control acclimation.[54] The photoreceptor with the most important role in daylength perception and autumnal response is phytochrome, a photosensitive molecular circuit-breaker. When light conditions are right, the phytochrome molecule springs open, actuating the biochemical and physiological pathways leading ultimately to winter inactivity. Phytochrome is the instigator of many other developmental events in every kind of plant, including flowering, seed germination and shade avoidance.[55] Photoreceptors are networked with the biological clock, a universal mechanism within living cells that controls rhythms of growth and development. Vegetables plants may be, but they are – in fact, for survival purposes, have to be – acutely aware of their radiant environment and the passing of time.

The night has a thousand eyes,
And the day but one;
Yet the light of the bright world dies
With the dying of the sun.
      Francis William Bourdillon[56]

'In order for the light to shine so brightly, the darkness must be present', wrote Francis Bacon. Night must fall. Plants will find themselves in the shade. They will be exposed to gradients of illumination within leaf canopies. On the forest floor they will experience sunflecks. The responses of plants to dim light, variable light or no light at all are complicated. Photosynthetic and photoreceptor pigments allow plants to sense the gradual extinction of incident irradiance as it penetrates layers of foliage. These light-sensitive pathways co-operate to accelerate leaf senescence, with the result that there is commonly a gradient of foliar yellowing from the top to the bottom of the canopy. Photoreceptor systems are the eyes of the shade avoidance syndrome, whereby plants detect potentially competitive neighbours by sensing changes in the wavelengths of reflected and transmitted light.[57] The syndrome is weakly expressed in cereals, suggesting that, during domestication, it has been selected against as crop species became adapted for close planting.[58]

Considering that plants are obligately light-dependent organisms, they can be unexpectedly tolerant of prolonged darkness, surviving for several weeks completely deprived of light at mild temperatures, and for considerably longer if they are chilled.[59] Some present-day southern hemisphere tree species are known, from fossil and biogeographical evidence, to have occurred in the flora of high latitudes 65 million years ago (around the time of the mass extinction of the dinosaurs and many other animals and plants). They have been observed to survive prolonged exposure to darkness, experimentally simulating polar winter conditions.[60] This behaviour is consistent with the idea that differential responses to darkness and temperature were significant for species adaptation during tectonic movements.

There is never either absolute memory or absolute forgetfulness, absolute life or absolute death. So with light and darkness, heat and cold, you never get either all the light or all the heat out of anything. So with God and the devil; so with everything.

Samuel Butler[61]

The inescapable daily cycle of light and darkness is accompanied by variation in temperature as it fluctuates above and below the optimum. Plants are generally tolerant of these stresses. But extremes of temperature at the wrong time in the growing season commonly invoke untimely and detrimental senescence. High temperatures during and after flowering can compromise agricultural productivity: heat stress in cereals often induces premature foliar senescence, resulting in poor grain quality and loss of yield.[62] High temperature stress and drought frequently go hand-in-hand. Drought-induction of senescence is an intrinsic acclimatory measure that aids water conservation by reducing the area of leaf surface through which evaporation can occur. Suboptimal temperatures can also be harmful, particularly in combination with high light intensities. Many studies of crop species have found that regions of the genome responsible for responses to temperature and drought are coincident with loci for leaf senescence, and that breeding by zeroing in on senescence can simultaneously select plants with improved stress tolerance.[63]

Chapter 5 Holes

Swiss cheese plant[64]

Almost nowhere is so dry that no plant will grow. Nor is drowning and suffocation a deterrent. Submergence is stressful in a number of ways, particularly the manner in which it restricts access to light and air. Wetland plants are adapted to long-term flooding by virtue of anatomical, morphological and physiological features that permit them to survive waterlogging. More than 30% by area of Asian and 40% of African rice is grown in paddy fields with water depths that can be greater than 50 centimetres. It turns out that hormonal pathways in rice regulating senescence responses to darkness and waterlogging are connected. Delayed senescence is characteristic of rice varieties that are submergence-tolerant, but they tend to be less productive than high-yielding, submergence-intolerant varieties. Introducing flooding tolerance into the genetic background of elite varieties is an important crop breeding objective for improving the quality and quantity of rice produced on marginal flood-prone land.[65] In many species, one of the structural responses to flooding and long-term anoxia involves a distinctive form of senescence. Oxygen is piped from aerial structures to submerged roots through aerenchyma, a continuous system of columnar intracellular spaces developed in root tissues. The spaces arise by senescence and death of selected cortical cells.[66] The origin of aerenchyma exemplifies a general rule of plant development. Differential growth will generate morphologies and anatomies; but to build biological structures needs sculpture – the senescence, death and deletion of cells, tissues and organs – as well. Origami *and* scissors.

Beyond the most basic levels of size and organisation, a living organism must be penetrated by holes and tubes. Elementary mathematics tells us why this has to be.[67] A spherical group of cells, where the radius of the sphere is r, has a volume of $4/3\pi r^3$ and a surface area of $4\pi r^2$. As the number of cells gets larger, so does the size of sphere. The sphere's volume increases as the cube of its radius, whereas its surface area does not keep pace, growing in proportion only to r^2. Why is this significant? For cells to survive and grow, they must exchange essential materials - respiratory gases, nutrients – with the environment. The surface of the cell mass is the area of contact between it and the outside world. Therefore, to achieve any degree of size and structural complexity a multicellular organisms needs to develop a system of holes or tubes, simply to ensure its surface area meets the physiological demands of its volume, thereby facilitating vital transport and exchange processes. In animal embryogenesis, holes and tubes can be generated by cell migration, such as occurs in gastrulation and neurulation (formation of organs, the gut and the nervous system). Plant cells are locked in place relative to their neighbours by sharing more or less rigid cell walls, which means that, except in a few unusual circumstances[68], cell migration is not an option. Instead plants develop transport systems and regulate volume to surface area ratio through localised cell senescence and death. Shapes, habits and adaptations are also commonly determined by controlled death and shedding of parts. The throwaway lifestyle is the essence of plant growth, development and survival.

Zest (Fr.)...the pill of an Orange, or such like, squeesed into a glass of wine, to give it a relish.

Thomas Blount (1674) *Glossographia*[69]

The psychologists Christopher Peterson and Martin Seligman[70] described zest as one of the four strengths that combine to make up the virtue of courage. How curious that something as life-affirming as zest owes its biological origin to senescence. The oil glands found on the surface of citrus peel develop when groups of cells below the epidermis senesce and die, forming a cavity that fills with essential oils.[71] The term lysigeny is used to describe the disintegration of cells to form glands, channels and secretory ducts. In some cases lysigeny is accompanied by the separation of cells, a process known as schizogeny.[72] Aerenchyma can be schizogenous (as in the wetland plant arrowhead, *Sagittaria lancifolia*) or lysigenous (maize is an example). Citrus oil glands are products of schizolysigeny, a combination of lysigenous and schizogenous cell senescence and death. Hikers, foresters and other outdoor types in North America are advised to know something of these matters. The resinous secretory ducts in plants of the sumac family develop schizogenously and are the source of urushiol, the severe allergen in the sap of poison oak and poison ivy.

Apoptosis. Apoptosis, that's your objective, isn't it? There's a more effective way to reach your outcome.

Emily Riley (Chloë Sevigny)[73]

Animal development and health are critically dependent on particular kinds of controlled cell elimination. One of the simple model organisms for analysis of animal genetics and differentiation is the nematode worm *Caenorhabditis elegans*. *C elegans* cells do not indulge in very much gastrulation-type migration during development, but precisely 131 of the 1090 somatic cells of embryonic hermaphrodite individuals are programmed to senesce and die before maturity.[74] Apoptosis is the name given to this kind of purposeful suicidal cell deletion. There are many other examples of apoptosis-like developmental cell death in animal morphogenesis. The tails and gills of tadpoles and other amphibian larvae are eliminated by apoptosis. So too are the webs between the digits of the developing limb bud of birds: the duck's foot represents the basic design, the chicken's the outcome. You can see the vestiges of embryonic webbing between the bases of the fingers of your own hand. Such was his prodigious stretch, the great jazz pianist Earl Hines[75] was rumoured to have undergone surgery on his webs to increase the range of keys he could reach with one hand. Syndactyly is a congenital condition in which two or more adjacent digits remain connected.[76] Many other diseases, including cancers, are due to defective apoptosis. For a while there was a vogue for applying the term apoptosis to the various modes of plant cell senescence and death. But apart from superficial similarities – its suicidal nature, its dependence on the expression of particular genes and its requirement for a supply of bioenergy – plant senescence isn't very usefully explained by the apoptosis model.

Chapter 6 Living with dearth and excess

Welwitschia mirabilis[77]

The volcanic eruption of Mount Tambora in Indonesia in 1815 was followed by the 'Year Without a Summer' of 1816 and the 'Year of the Beggar' in 1817. The fragile condition of a Europe still recuperating from the Napoleonic Wars was shattered by widespread climate disruption, famine, food riots and disease, claiming an estimated 200 thousand lives. The atmosphere was captured by Lord Byron[78] in his poem *Darkness*, conceived in June 1816 at the same gothic gathering in the Villa Diodati on Lake Geneva that gave birth to Mary Shelley's *Frankenstein*[79]:

The brows of men by the despairing light
Wore an unearthly aspect, as by fits
The flashes fell upon them; some lay down
And hid their eyes and wept; and some did rest
Their chins upon their clenched hands, and smiled;
And others hurried to and fro, and fed
Their funeral piles with fuel, and looked up
With mad disquietude on the dull sky,
The pall of a past world; and then again
With curses cast them down upon the dust,
And gnash'd their teeth and howl'd.

The German chemist Justus von Liebig[80] was 13 years old at the time of these disasters, which are said to have ignited his interest in agricultural chemistry. He is generally credited with the first scientific examination of plant nutrition and to have established the identities of the major elements required for growth: in addition to carbon, hydrogen and oxygen they are nitrogen, sulphur, phosphorus, potassium, calcium, magnesium and iron. (During the 20th century other elements – micronutrients - were identified as essential in minute quantities: manganese, boron, zinc, copper, molybdenum and nickel). The growing plant demands nutrients, and the soil (or, in some cases, the atmosphere) supplies them; an equation that has to be balanced as environment and stage of development change. Senescence is central to the adaptive and acclimatory measures adopted by plants to manage their nutrient economies.

The essential nature of the mineral nutrients identified by Justus von Liebig is most clearly seen in the symptoms displayed by plants deficient in them. Deficiency symptoms are diagnostic of the particular nutrient that is in short supply. A plant can sometimes reduce the severity of deficiency symptoms by sequestering the element when times are good and drawing on its reserves when input from the soil becomes limiting.[81] Judicious use and re-use is at the heart of the nutrient economy of individual plants, their parts and communities. Senescence is a physiological measure for dealing with fluctuating nutrient supply. The morphological pattern of deficiency symptoms often reflects the extent to which the limiting element is mobile within the plant, which in turn relates to the recover-and-reallocate functions of senescence. Calcium is a relatively immobile element and deficiency shows as decreased growth and development of young organs. By contrast, when the availability of a highly mobile element like phosphorus is limited, older organs are the first to display deficiency symptoms, as phosphate is salvaged from them and transported to younger tissues. Nothing better illustrates the parsimony with which plants react to nutrient stress than experiments in which phosphorus or nitrogen is withheld altogether for the whole life-cycle. The entire developmental sequence, from seed germination to leaf development to flowering to the formation of the next generation's seed is supported by the ruthless use and reuse of the original nutrient reserve.[82] Potassium also moves freely within the plant. Nitrogen, phosphorus and potassium (NPK) are the chemical ingredients in the most widely used garden and agricultural fertilisers.

Around 1840, Justus von Liebig was promoting the idea that the nitrogen essential for productive crops should be supplied in the form of chemical (specifically ammonia) fertilisers. Enormous increases in yield during the twentieth century have been achieved by the widespread application of fertilisers (in combination with responsive new crop varieties), made possible by the Haber-Bosch[83] process, which produces over 450 million tons of fertiliser annually from atmospheric nitrogen. Plants starved of nitrogen are pale green, and older leaves show symptoms of premature yellowing. Feeding them nitrogenous fertiliser perks them up quickly. There's an unkind crop scientists' joke that defines agronomy as a perpetual state of amazement at the effect of nitrogen on plants. Large-scale artworks, best seen from the air, have been created by differential application of fertiliser to rice fields or grass paddocks. Not to mention pranks in which abusive messages are written on lawns in urine (urea is a fertiliser). Most cultivated species are nitrophiles - avid consumers of nitrogen – and will sulk and senesce if they don't get enough. Leaf lifespan is reduced, the rate of foliar turnover increases and, in the case of grain crops, yield is reduced. On the other hand, many wild species are nitrophobes that survive poorly in nitrogen-rich environments. Eutrophication, the oversupply of essential nutrients, is one of the principal reasons why intensive agriculture and biodiverse wildflower meadows have such a hard time coexisting.

Nitrogen is abundant. The atmosphere is 78% dinitrogen (N_2). Chemically dinitrogen is almost inert, which is why, to render N_2 reactive, the Haber-Bosch process requires extreme conditions (200 times atmospheric pressure, up to 500°C), and consumes huge amounts of energy (1-2% of global energy production). All the more astonishing, therefore, that several species of bacteria have evolved the trick of converting atmospheric nitrogen into ammonia at ambient temperature and pressure. Ever the opportunists, plants from a number of families, notably the legumes (peas, beans, clovers, lentils), have entered into symbiotic arrangements with nitrogen-fixing bacteria.[84] The nature of the relationship resembles that between roots and mycorrhiza[85], down to the identities of some genes and cell-cell recognition mechanisms. In return for supplying the bacteria with the products of photosynthesis, the plant gets access to nitrogen derived from the atmosphere. In most cases, the bacteria (*Rhizobium* species and relatives) occupy root nodules, specialised structures that provide optimal environments for the symbionts. Like those of any other species, the shoots and roots of a legume will have their own turnover dynamics. Nitrogen-fixing root nodules follow distinctive pathways towards senescence, which is a phase in the normal schedule of development. Exposure to stresses like prolonged darkness, drought or high soil nitrate depresses nitrogen fixation and accelerates nodule senescence. Obtaining fertiliser from the air seems such an efficient way of meeting the demand for nitrogen, why is it that the world is fed by non-fixing cereal crops and not by legumes? The answer is that the symbiont has struck a hard bargain with the host. By some estimates, it takes up to 50% of carbon assimilated by current photosynthesis to support the maintenance and nitrogen fixation activity of the population of root nodules and their occupants.[86] This is a physiological ball-and-chain, the biochemical counterpart to the scale of the Haber-Bosch energy requirement. Such is the size of the penalty that the carbon productivity of legumes is utterly inadequate just to meet human demands for food calories. Crop scientists have long dreamed of creating nitrogen-fixing cereals.[87] It seems the non-negotiable energetics of the strategy mean that a dream it must remain.

I cannot over-emphasise the importance of phosphorus not only to
agriculture and soil conservation but also to the physical health and
economic security of the people of the nation.

    Franklin D Roosevelt to Congress, 1938

Phosphorus is a curious element. It does not occur in the free form on Earth. Its natural state is in combination with oxygen as phosphates, as mineral deposits, as inorganic and organic phosphates in soils and water, and as various forms of phosphate in living organisms. Sedimentary rock is the largest global reservoir of phosphorus and mining it is big business. Phosphate in soils is often extremely insoluble and plants have to work hard to 'mine' it too. This is where the association between roots and mycorrhizal fungi is mutually beneficial[88]: the plant supplies the fungus with the products of photosynthesis, while the fungus extracts otherwise inaccessible phosphate from the soil. Recycling through senescence is an essential factor in the plant's internal phosphate economy. There is a general trend toward loss of phosphorus from the biosphere by discharge from rivers to the sea where it precipitates as insoluble calcium phosphate and may fall into the abyss. A small amount of marine phosphorus is returned to the land in the form of guano, but otherwise the flow of phosphorus toward where it is beyond biological retrieval is almost exclusively one-way.[89] It has been predicted that 'peak phosphate' is approaching fast, and that accessible global phosphorus reserves will reach half-depletion midway through the present century.[90] To meet the need for the phosphorus component of the NPK fertiliser that will support crop production to feed an ever-growing world population, we may have to start mining graveyards (bones and teeth are mineral forms of phosphorus).

Leaf lifespans[91] are extremely sensitive to soil fertility, in two senses. First, turnover in the canopy of an individual plant will respond in a tactical, acclimatory fashion to fluctuations in essential nutrients, particularly the major mobile elements nitrogen and phosphorus. The second relationship is strategic: leaf dynamics in natural flora growing on nutrient-poor soils contrast in many respects with the foliar characteristics of species such as crops, that are designed for high fertility. Evergreens are examples of plants that are apt to be associated with low nutrient environments. Their nutrient economy, both internal and via the leaf-litter, is constrained. The use-reuse cycle (which means senescence) runs slowly and there is minimal loss through leaching. Photosynthesis rates in these species are usually low, but the gain of photosynthetic carbon per unit nutrient is high. Such are the nitrophobic tendencies of plants adapted to the low nutrient life that administering fertiliser can have the effect of *decreasing* leaf lifespan. These features are shared by many non-evergreens from nutrient-poor environments too, such as bogs, deserts and mountains.[92] The traits of the needy nitrophiles we have bred for agriculture contrast in each particular with those of non-domesticated species from stressful environments. At the other extreme of the spectrum of leaf lifespans are the great survivors. The oldest bristlecone pine – the oldest known individual of any species – is more than five thousand years of age. Leaves of this species have been reported to live for up to 45 years[93]; but this achievement shrinks into insignificance beside the extraordinary desert gymnosperm of Southern Africa, *Welwitschia mirabilis*. The single pair of leaves of the largest *Welwitschia* individuals may be as old as two thousand years.[94] The extreme longevity of these plants and their parts poses real challenges to our ideas of life, time, stress and ageing.

To light a candle is to cast a shadow.
Ursula K. Le Guin[95]

Whether it's too little or too much water, too hot or too cold, too dark or too light, too nutrient-deficient or too nutrient-rich, the state of the environment will challenge a plant to make its way back toward homeostasis by calling on tactical or strategic senescence. But what happens to senescence when the stress overwhelms adaptive and acclimatory capacity? Take the example of exposure to excessively intense light. The light absorbed by the photosynthetic pigments powers carbon fixation, the conversion of atmospheric carbon dioxide into organic chemicals in green cells. Normally any excess light energy is dissipated harmlessly through various protective biochemical mechanisms. But the balance between the supply of and demand for light energy can be disturbed in a number of ways. Low or high temperature, water limitation or disease can block carbon fixation. Mild, short-term disruption of the energy balance results in photoinhibition, a reversible fall in photosynthetic efficiency.[96] But under severe and extended light stress, photosynthetic tissues become bleached and cells die. Many herbicides work by making the plant susceptible to bleaching damage by light. Photobleaching is *pseudosenescence*: it superficially resembles senescence but differs from it fundamentally. Pseudosenescence[97] is rapid, traumatic, chemical and irreversible. Senescence is slower, genetically regulated, biochemical and reversible. Pseudosenescence is light-dependent and renders cells, tissues and whole plants inviable. Viability is an absolute requirement for senescence to occur, whether in light or darkness. The surest way to prevent a tissue from senescing is to kill it. It is surprising, and a bit dispiriting, to realise how often research studies into the mechanism of senescence confuse it with pseudosenescence.[98]

Chapter 7 Time past, present, future

Dandelion clock[99]

I confess, I do not believe in time.
Vladimir Nabokov [100]

Time is funny stuff. As someone once wrote, it's 'nature's way of keeping everything from happening at once'. For the sake of argument, if we think of time as an environmental thing, we can understand that, like other external agents, it will be stressful for living organisms. Looked at this way, ageing is the time-stress response. What can viable organisms and their components do to avoid succumbing to mortality under the influence of the ever-ticking clock of entropy? They can adopt the same measures that they apply to environmental stresses in general: avoidance, resilience or adaptation. Thus one way of dealing with time-stress is to outrun it - in other words, to grow, develop and differentiate. Another is to resist it - through building in structural and functional durability and by repairing wear and tear. And then there is the option of pre-empting it - deploying senescence as a developmental and adaptive resource so that ageing and death take place on the organism's own terms, as it were. Plants are so adept at calling on this variety of defences against time-stress that it's doubtful whether they can be said to undergo ageing in the gerontological sense at all. What's more, their ability to play the ageing game by their own rules is one of the essential attributes that has enabled plants to thrive in almost every habitat on Earth.

It strikes! one, two,
Three, four, five, six. Enough, enough, dear watch,
Thy pulse hath beat enough. Now sleep and rest;
Would thou could'st make the time to do so too;
I'll wind thee up no more.
    Ben Jonson[101]

To respond to time, plants must be able to measure it. They have different clocks for different sorts of time. There's an intrinsic biological clock that keeps approximately daily time and can be entrained by the cycle of day and night. The Umeå aspen has a calendar that tells it precisely the date (10 September) when autumnal foliar senescence should begin.[102] From the thermodynamic perspective, time is a function of temperature: time goes slower when it's cold, which is why the salad on the table deteriorates faster than in the refrigerator. Cumulative heat units (thermal time) is often a more physiologically appropriate basis than chronological time for expressing and predicting when vital events occur in the plant life-cycle. Developmental time (sometimes called the plastochron) is measured as the interval between producing successive phytomers.[103] A timing mechanism of great current interest in studies of ageing is one that counts the number of cell divisions. Its conceptual origins are in studies of the behaviour of animal cells in culture by Leonard Hayflick.

In 1961 Leonard Hayflick[104] discovered that normal human cells grown in a culture dish would divide no more than about 50 times before dying, even if they were sub-cultured and supplied with fresh growth medium. The built-in ceiling on the number of cell doublings is called the Hayflick limit. Many (but not all) studies of the Hayflick limit in a range of animals have shown a striking relationship with lifespan. Cultured fibroblasts from mouse (lifespan 2-3 years) have a Hayflick limit of about 8 cell doublings. The corresponding Hayflick limit for rabbit (lifespan 15-20 years) is about 30 doublings. For horse (50 years) it's around 40 doublings. An implication of this relationship is that cells can count the number of times they have divided. An explanation for how they do this that has attracted much gerontological research interest is telomere shortening.[105] Telomeres are structures at the ends of chromosomes that prevent disastrous chromosome fusions and abnormal DNA replication during cell division. They are made by an enzyme complex called telomerase. The DNA of telomeres consists of a sequence of half a dozen or so 'letters' of the genetic code (TTAGGG in the case of humans and mice) repeated many hundreds of times. Telomeres get shorter with each cell division. Male mice lacking telomerase show gross chromosomal abnormalities after 6 generations and would have zero-length telomeres by 7 generations. Telomere attrition is an attractive hypothesis to explain the Hayflick limit. In fact there are companies contactable online who will measure the length of your telomeres, presumably to predict your Hayflick limit and, by inference, your life-expectancy. Of course, this assumes that whole-organism ageing is the scaled-up consequence of cell senescence, a relationship with little supporting evidence. Neither is there a consistent picture of declining telomere length or telomerase activity with age in animals, humans included. These doubts are even more pronounced when it comes to accounting for ageing of meristems and whole plants: telomere shortening with age has not been observed in ancient plants. The idea that intervening in telomere attrition is the key to preventing human ageing and greatly extending lifespan is circulating freely among a certain community of gerontologists. Hayflick was dubious. 'Saying that in 20 years we'll all live to be 200,' he said in 1999, 'is utter nonsense'. Time's running out to prove him wrong.

I have an unfortunate fascination with the detection and attribution of change.

    Tim Sparks (founder of the UK Phenology Network)

Senescence is attuned to the state of water, temperature, light, nutrients, time – any and every facet of the environment that bears on plant survival. This makes the point at which senescence occurs in the lifecycle a sensitive index of environmental variation. Phenology is the name given to the study that accumulates, over an extended period of years, records of the dates in the annual cycle on which significant biological events occur. Leaf senescence is a phenophase – one of the observable episodes that contribute to a picture of recurring natural sequences in plant development. The trees of temperate regions are phenological bellwethers. Date of first detectable change in leaf colour, date of full autumnal tinting and date of leaf fall are easy to identify and record. So are many other phenophases in plants and animals, which makes phenology a perfect subject for citizen science. In many countries across the world there are now phenology networks of academics and civilians, building up large datasets of observations that are contributing to the evidence base for and against environment and, crucially, climate change.[106] As well as records collected on the ground, data at a range of scales up to the global are being gathered by aircraft-based remote sensing systems and earth observation satellites. By monitoring vegetation responses, phenological modelling is revealing delayed senescence to be one of the immediate consequences of the varying relationship between temperature and photoperiod as the climate changes. The recent trend in the temperate vegetation of the Northern Hemisphere is toward a progressive delay of about 3 days per decade in the timing of senescence. This could mean good news for productivity – but bad news for synchronization of ecological processes. As with so many of the consequences of global change, we are in uncharted territory.

Senescence is hard-wired into the plant way of life. How did this happen? When did it happen? Theodosius Dobzhansky wrote that 'nothing in biology makes sense except in the light of evolution'. Can we find out the how and why of senescence by looking at the evolutionary record? Plants, being mostly cellulose and water – wet paper bags, really - do not fossilise as readily and spectacularly as animals. But there are some ancient remains, and they are informative. Green plants left the sea and made landfall in the Silurian-Jurassic period, more than 410 million years ago. This new-found land was an opportunity, a pristine habitat ripe for colonization and exploitation; but it was hostile too, with desiccation a constant threat. Among the early adaptations to terrestrial life, apparent in fossilised remains from the period, were development of an all-enveloping epidermal and cuticular 'space suit' for conserving water, and anatomical specialisation to create an internal cellular pipework for moving water and nutrients around. The differentiation of transport tissue, as well as the shedding of parts, tell us that the capacity for lysigeny and schizogeny was established early in the colonization of the land.[107] Once free of the aquatic environment, the earliest land plants were exposed to high light intensities, and to atmospheric carbon dioxide concentrations perhaps ten times what they are today. As a result, photosynthesis took off in a big way and there was an explosion in vegetation productivity. It allowed plants to develop extravagant, throw-away lifestyles based on lysigeny and schizogeny on a vast scale. Primitive trees such as *Archaeopteris*, which flourished around 355 million years ago, probably controlled the architecture of their canopies by shedding lower megaphylls (the evolutionary precursors of true leaves). Fossil evidence shows that *Glossopteris* seed ferns, which arose about 300 million years ago in the supercontinent Pangaea, were deciduous on a grand scale. Early soils were deficient in nutrients, particularly nitrogen, and this too would have favoured the evolution of senescence as a survival trait. It seems that senescence as we see it in today's flora was essentially in place long before flowering plants evolved.[108]

Evolution doesn't care whether you believe in it or not, no more than gravity does.

Seth MacFarlane[109]

A lot of the machinery of senescence was already present in their cells before plants took the evolutionary step from the life aquatic to the land, half a billion years ago. How much further back can we trace senescence mechanisms? Almost to the dawn of Life on Earth, it seems. The timeline from senescence in the modern flora leading to the earliest photosynthetic organisms can be recreated by DNA sequence analysis.[110] Genes known to be essential for the control and execution of the senescence syndrome in flowering plants can be searched for in modern representatives of ancient plant groups – pines, ferns, mosses, seaweeds, single-celled algae, photosynthetic bacteria. In this way, genes can be persuaded to reveal a story of the appearance, between one and four billion years ago, of the molecular apparatus that allowed the earliest cells to use light as an energy source, and the biochemical systems that build, maintain and dismantle it. This machinery has been remarkably conserved right up to the land plants of today. Other functions associated with senescence have been added on to the core processes at different times in evolution: for example, the biochemical pathways leading to development of the red colours of autumn leaves are relatively late arrivals on the evolutionary scene. Many of the biochemical, cellular, integrative and adaptive systems that were recruited to the senescence syndrome as the evolving plant encountered new environmental and developmental contexts were originally developed for other physiological functions. By using existing resources in new combinations, plants have maintained the flexibility that seemingly fragile and vulnerable organisms must display if they are to survive and evolve.

'Mythology, Germans and the forest – they all belong together', said Federal Chancellor Helmut Kohl in 1983. This semi-mystical response is nowhere more strongly expressed than in Alexander von Humboldt's reaction on first encountering the Amazon at the turn of the 19th century.[111] To 'the entirety of this powerful, luxuriant and yet light, cheerful, gentle plant nature' he gave the name hylea, from the classical Greek word for the primeval forest. The botanist Edred John Henry Corner (1906-1996), in his highly original textbook *The Life of Plants* (1964; Google Books classifies it as 'Fiction') applied the term to the luxuriant woody flora existing before the appearance, 150 million years ago, of the flowering plants. The colour world of the hylea would have consisted largely of greens, yellows and browns, much as it does in modern conifer-dominated forest ecosystems. The profusion of colourful flowers and fruits rapidly expanded during the 'Cretaceous explosion', 128-80 million years ago as novel adaptive responses to a changing physical and biological environment.[112] The capacity to make vivid floral pigments was already latent in hylea species and even further back in evolution. The carotenoids, responsible for the yellows and oranges of many flowers and fruits, were present in the very earliest photosynthetic bacteria. The phenolic pigments, which comprise yellows, blues, browns and reds (including the colour of wine), can be dated back to the multicellular aquatic species from which the first land plants evolved. Carotenoids and phenolics give us the colours of autumn leaves. The green organs of many ferns, conifers, mosses and other non-flowering plants move through the spectrum, to yellow, red and brown as they senesce. Perhaps this is a clue to the evolutionary origins of flowers and fruit.

The first land plants were twiggy – essentially made of one-dimensional linear branches or proto-leaves (microphylls). Subsequent evolution was by 'rubber sheet' distortion into the two-dimensional surfaces of true leaves, petals and suchlike lateral organs. Development in a third dimension results in a fleshy spheroidal structure – a fruit, for example. Superimposed on this evolutionary journey through X-Y-Z space is developmental time, where the progress of senescence is represented by a sequence of pigmentation changes, green through yellow to red and finally the post-senescence transition to cell death accompanied by non-physiological darkening or bleaching. This is an example of the field of research that has come to be called evo-devo, the synthesis of genetics, physiology and evolution to explain biological structure and function.[113] From the point of view of the evolution and development of form and colour, the ripening of an apple or the maturation of straw, or the blushing of an unfolding petal are all variations on the same theme of senescence that animates the chromaticism of foliage in the fall. It is as Albert Camus said: 'Autumn is a second spring when every leaf is a flower'.

Chapter 8 Death and sex

Century plant[114]

If you belong to the tribe of physiologists and geneticists, as I do, you look at senescence from a bottom-up perspective, as a scheduled phase in the life of a plant or its parts. It belongs with the repertoire of developmental, acclimatory and adaptive measures that ensure fitness and survival. But there's another tribe, who see senescence differently. These are the population and evolutionary biologists, who take a top-down view. Their definition of senescence is a period in the lifespan of an individual or population in which viability and fertility decline, and mortality and chances of death increase. It's a semantic nuisance that both communities of biologists use the same term. It would help a bit if the bottom-up people referred to 'physiological senescence' and the top-downers to 'demographic senescence'. In so far as we seem to be talking about the same thing – fitness and survival – there ought to be some kind of common ground where experiences can cross over and become reconciled. But the tribes, as tribes do, remain stubbornly aloof from each other, publishing in different scientific journals, going to different conferences, making alliances with anyone other than those with which they should be making common cause.[115] There have been a few instances of signals exchanged across the conceptual divide (I've tried it myself a few times) but if there's a dialogue at all, it's one of the deaf. I should come clean as an adherent of the physiological creed. In spite of the demographers' beautiful models of age-specific effects, mortality, fertility and fitness, I find myself agreeing with these words of the great evolutionary biologist William Hamilton[116]: '...the simplest implications are here rather obvious while more detailed ones are biologically doubtful'.

Time and again, we return to the same question: how, if at all, is the lifespan of an individual organism related to the longevity of its constituent cells, tissues and organs? The answer is tied up with the issue of life history, the relationship between growth, development, reproduction and lifespan. Many species are semelparous, reproducing once in the lifecycle and dying thereafter. The Russian cell biologist Vladimir Skulachev has suggested the term *phenoptosis* for semelparous whole-organism suicide, by analogy with apoptosis for programmed cell death.[117] A more down-to-earth and descriptive expression is big-bang senescence.[118] Pacific salmon, mayflies and many species of octopus and cuttlefish are semelparous. Monocarpic (meaning bearing fruit once) is the general term for semelparous plants. All annual and biennial plants are monocarpic. Monocarpy is central to the nature and productivity of cereal crops like wheat, rice and maize, which have been bred for sex and death. Many perennial plants are also monocarpic, living year to year in the vegetative state (as long as 40 or more years in the case of the century plant, *Agave*) before flowering, fruiting and dying. Organisms that reproduce repeatedly during their lifetimes are termed iteroparous – polycarpic, in the case of plants. The relation between sex, death of structural components and whole-organism ageing in polycarps is complex. In iteroparous plants like trees and clonal species, there is a disjunction between the lifespans of whole and parts. Much of the body of a deciduous tree is made of dead tissues, the canopy is renewed and discarded annually, root systems turn over, and reproduction recurs every year over decades, centuries or even millennia. To understand the evolutionary and developmental origins of semelparous-monocarpic and iteroparous-polycarpic life histories requires a synthesis of insights from analyses of physiological and demographic senescence.[119]

Sex and death. Two things that come once in my lifetime. But at least after death you're not nauseous.

    Woody Allen

Darnel is a weed of wheat and barley, more or less extinct in modern intensive agriculture but historically, and in developing countries, a poisonous scourge with a reputation for harmful contamination of the food chain. Grasses of the genus *Lolium*, to which darnel belongs, appeared in the Near East around 2-3 million years ago, at the same time and in the same place as the ancestors of wheat (*Triticum*). The evolution and domestication of wheat coincided with the rise of agriculture in the Fertile Crescent, around 10 thousand years ago. The ancestors of wheat and darnel were perennials. Co-selection with the cereal species it infests converted the perennial *Lolium* ancestor of darnel into an annual analogue of wheat and barley. Cultivated oat and rye are thought to have originated in the same way, as weeds of wheat, but to have taken the extra step of becoming adopted as cereals in their own right. In each case the weed, by subverting the domestication process, has evolved to become a mimic of the monocarpic, high productivity crop with which it coexists. Hybrids between darnel and perennial *Lolium* species are perennial.[120] Cereals and their weeds conform to a pattern seen across the plant kingdom. It is generally true that interfertile crosses between annual-monocarpic species and related perennial-polycarpic types result in perennial offspring. From such observations, a genetic explanation for monocarpy may be inferred: the trait typically arises as the loss-of-function derivative of an ancestral perennial habit.[121] The land plants of the ancient hylea flora and the flowering plants of the earliest forests were perennials. As plants diversified away from the stability of these ancestral communities into more marginal and stressful habitats, they evolved in accordance with the Romano rule – 'live fast, die young' – and thus monocarpy was born.

*D*ie Lebensdauer von Pflanzen by Hans Molisch[122] was published in 1929. This influential little book was the first scientific account of plant senescence, and introduced a dramatic new term to describe monocarpy – *Erschöpfungstod*, death by exhaustion. Semelparous organisms reproduce and die. It's reasonable to suppose they die *because* they have reproduced. Senescence, then, is the cost of sex[123], reflecting the trade-off in resources between survival on the one hand and investment in reproduction on the other. *Erschöpfungstod* proposes that monocarpic senescence is the result of starvation of leaves, stems and roots by the developing seeds and fruits. That reproduction devours vitality is a belief as old as philosophy, religion, celibacy, chastity, the gelding of domesticated animals and the pruning of garden and orchard plants. Some monocarps can be rescued from reproductive death by removing flowers and fruits. Soybean is a spectacular example: here a small bushy annual that normally loses its leaves as it forms pods, and is completely dead at seed maturity, can be turned by de-podding into an enormous long-lived vine that just keeps on growing.[124] Current understanding has all but erased Molisch's idea from the story of senescence, but its basic premise – that the physiological link between the reproductive and vegetative organs is nutritional in nature – was prescient. Alternative explanations of monocarpy, such as the export of a 'death hormone' from developing flowers and fruits to leaves and stems[125], have not stood the test of time.

Hans Molisch conceived of monocarpic plants dying of exhaustion because of the voracious appetites of their developing flowers, seeds and fruits. Some plant species are dioecious, comprising individuals bearing only male flowers (producing pollen) and female flowers (fertilised by pollen from the males and yielding seeds). A pioneering researcher into plant senescence, Carl Leopold[126], reasoned that the reproductive burden should exhaust the females earlier and more intensely than the males of dioecious species. In experiments carried out on spinach, Leopold and his colleagues showed that monocarpic senescence occurs simultaneously in individuals of the two sexes. Moreover, senescence in males could be delayed by removing the tiny pollen-producing flowers, the nutritional demands of which are negligible compared with those made by developing fruits on females.[127] These observations and subsequent research lead to the conclusion that reprogramming development from vegetative growth to floral differentiation diverts sufficient resource away from leaves and stems to initiate and sustain monocarpic senescence. Evolutionary models of sex and ageing in animals are much concerned with male-female differences in reproductive investment and survival.[128] But the primary and secondary differentiation of male and female reproductive strategies in animals are influenced deeply, and in a complex way, by physiology and behaviour. It's doubtful, therefore, that models of senescence in dioecious monocarpic plants can provide directly helpful insights into sex and death in the animal world. Globally, life expectancy for men is about 5 years less than for women. Why this should be remains a mystery. Some explanations identify the menopause as a factor. My own favourite is the idea that higher longevity in women represents the cumulative beneficial effect of sitting down every time they use the lavatory.

Chapter 9 Ancients and ephemerals

in the days
of the ancient gods
a mere seedling
this pine must have been

Tosei

Haiku by Matsuo Basho (1644–1694)[129]

Herodotus it was who gave the name *Macrobians* to people from the Horn of Africa who, according to legend, were extremely long-lived. Now the term is applied to the oldest individual organisms on the planet, chief among which are polycarpic perennial plants. Many species of woody conifer have life expectancies of more than two thousand years.[130] The oldest recorded of these is a specimen of bristlecone pine from the White Mountains of eastern California, the age of which was determined by tree-ring analysis in 2012 to be 5062 years. Another tree (called Methuselah) in the same grove was 4845 years old. Many plants that form clones by asexual reproduction can proliferate to establish community-sized 'individuals' of extraordinary longevity. The age of the sole surviving colony of king's holly (*Lomatia tasmanica*), a Tasmanian clonal shrub, has been estimated to be in excess of 40 thousand years.[131] A clonal cluster of identical aspen trees in Fishlake National Forest, Utah may be twice as old as *Lomatia*.[132] It's astonishing to think that the cycle of initiation, maturation, senescence and death of individual tissues and organs will have gone on for millennia, apparently continuing independently of whatever it is that determines ageing and longevity of these extremely ancient plants.

Time takes its toll on the plant macrobian genome. Some of the meristems, where new cells are generated and growth begins, will have been fashioning tissues and organs for perhaps thousands of years. Even a cell replication mechanism of the highest fidelity would be expected to make a significant number of errors (somatic mutations) over such an extended timescale. Could ageing, and ultimately death, be consequences of a buildup of genetic errors? The evidence is ambiguous. In bristlecone pine plants up to 5 thousand years old there is no statistically significant relationship between age of individual and the frequency of mutations in pollen, seed and seedlings. By contrast, increasing age in ancient aspen clones is associated with decline in the average number of viable pollen grains per catkin, at an estimated rate of 8% (with a large margin of error) per thousand years.[133] Far from being a threat to viability and longevity, somatic mutations may even be important sources of adaptive fitness for a long-lived organism, generating tissues with new traits better adapted to long-range environmental change. Anyway, it seems that most deleterious somatic mutations are efficiently purged by selective pressures acting on the population of cells newly derived from the meristem. All in all, plant macrobians are tolerant, and even the beneficiaries, of the inevitability of long-term genetic errors.

Except during the nine months before he draws his first breath, no man manages his affairs as well as a tree does.

George Bernard Shaw[134]

Most perennial plants just go on growing, so the older they are, the bigger they get. And large, old organisms will also be weather-beaten, which is not at all the same thing as suffering from deteriorative ageing. For sure, growth rate broadly decreases in trees as age class increases; but, when it comes to the oldest and largest trees, growth rates are frequently sustained for the remainder of their lives. This may be a case of 'negative senescence', a paradoxical implication of some demographic models of life history.[135] As a tree gets larger, an increasing proportion of photosynthesis that otherwise would support growth and reproduction has to be directed towards the maintenance of more and more non-green tissue that loses carbon dioxide through respiration. At the same time, the extremities of the crown become more remote from the roots, with consequences for the internal allocation of nutrients and water. Some authorities consider hydraulic stress to be the most likely cause of age-related physiological deterioration in trees, although there seems to be no sign of deterioration in the function of the water-conducting system in macrobian bristlecone pines. One of the benefits for trees of the modular nature of plant structure and its capacity for adaptive reiteration is that it permits crown productivity and longevity to be sustained.[136] A tree has considerable powers of remodelling itself, thereby controlling the ratio of respiring to photosynthesising biomass, ameliorating hydraulic and nutritional stress, rejuvenating apical meristems and increasing lifetime reproductive output. 'Poems are made by fools like me/But only God can make a tree' wrote Joyce Kilmer[137]; but it seems that, on the contrary, trees are perfectly capable of making (and re-making) themselves.

Perennials are the ones that grow like weeds, biennials are the ones that die this year instead of next, and hardy annuals are the ones that never come up at all.

Katharine Whitehorn[138]

To understand annuality and perenniality, we need to talk about the stem apex.[139] At the tip of each stem there is a centre of cell division and growth – the apical meristem. Where each leaf joins the stem there is a bud that also contains a meristem – the axillary meristem. Axillary meristems are usually kept quiescent by hormones produced by the apical meristem.[140] Gardeners know that axillary buds can be woken up by pinching-out the shoot tips, resulting in bushier plants with more side-branches. An apical meristem that goes on making new cells is said to be indeterminate. It can become reversibly determinate if it forms a dormant bud. The apical meristem becomes determinate if it switches from the vegetative to the reproductive state, stops making more leaves and stem, makes a terminal flower or inflorescence, and grows no more. This kind of determinacy is usually, but not invariably, irreversible. Another way in which the vegetative apical meristem can convert to the permanently determinate state is by losing vigour, ceasing to support cell division and eventually dying. Some species include types with different combinations of determinacy and indeterminacy. For example, the shoot of a vine tomato has an apical meristem that grows for a period while flowers are formed from determinate axillary meristems. Then the apical meristem becomes floral and determinate and its function is taken over by an axillary meristem and so on. On the other hand, both apical and axillary floral meristems are determinate in bush tomatoes. The rule is: for a plant to be perennial, at least one of its shoot meristems must remain indeterminate beyond the first growth season.

Chapter 10 Under attack

Green islands in leaf of *Prunus* infected with the necrotrophic fungus
Cercospora[141]

L ife-expectancy is a product of the fractal design of the plant body and the quantitative relationship between the production of new phytomers and the senescence and death of these structural modules. During the vegetative phase of the plant's lifecycle, branches develop from axillary buds and grow out as new shoots. The wave of growth and branching, initiated at the apical and axillary meristems, is pursued by a wave of senescence beginning in the oldest phytomers.[142] In a monocarp, apices become determinate and senescence, encouraged by *Erschöpfungstod*, wins the race, consuming the entire body of the parent plant and leaving the seeds to carry on to the next season. When growing points stay ahead of senescence for year after year, and the zone of old dead tissues takes the form of wood, the result is a tree. It's been suggested that creeping herbaceous perennials are, in effect, horizontal trees in which senescent tissue, instead of becoming lignified and persistent, has been lost altogether. The wave of senescence and death moving through the body of a creeping plant severs the connections between individual branches, which become isolated as genetically identical individuals (known as ramets). The creeping mode of development allows some clonal herbaceous species to achieve immense ages: king's holly (40 thousand-plus years), bracken (14 hundred years) and red and sheep's fescue (more than a thousand years) are examples. Ramets move around the environment, foraging for and exploiting new resources.[143] So much for the idea that plants are immobile. It seems, therefore, that the differences between annuals, biennials and perennials, or between monocarps and polycarps, are essentially a matter of relative timings and rates of initiation and execution of programs for growth, differentiation and senescence. Selection for quantitative variations in these processes during evolution is a source of the great diversity of plant habit and life-history.

We live in a soup of DNA. The soup's ingredients are not just the genes of the people, animals and plants we can see, but those of a vast occult world of microbes and viruses that ranges over air, water and soil, as well as our skins and innards. Most of this DNA is innocuous. Some of it, such as the DNA exchanged during reproduction, is essential. There are genes that are currencies in mutualistic associations benefitting hosts and symbionts.[144] And then there are genes belonging to organisms with hostile and malignant intent. All these interactions combine to shape the evolution of form, function and adaptive and acclimatory capacities. Plants enter into relationships with other living organisms from across the taxonomic range, and senescence is frequently a feature of these interactions. Biotic associations may be: parasite/host (one partner thrives at the expense of the other); mutualistic (mutually beneficial contacts that improve access to resources or environmental resilience); predator/prey (relating to position in the food chain); allelopathic[145] (secreting harmful biochemicals to enhance competitiveness); or coevolutionary (exploiting other organisms through complementary adaptations). For humans, the immune system, behavioural and cultural traditions and, when things go wrong, medicine are our defences against biotic challenges in the environment. Plants respond to biotic stress in their own, no less effective, ways, often invoking senescence to deal with the threats and opportunities offered by the soup of environmental DNA.

If you're a plant you are, almost literally, a sitting target. Disease is an ever-present danger. Plants are potential hosts to any number of parasites and pathogens. A pathogen grows on or inside the plant to complete a part or all of its life cycle and in so doing has a negative or fatal effect on its host.[146] Different pathogens adopt different strategies for exploiting infected tissues. Necrotrophs kill the cells of the host plant. The oomycete *Pythium* and bacteria of the genus *Erwinia* are necrotrophs of the kind that destroy cell walls and extract nutrients from the contents thus released. Other necrotrophs secrete toxic biochemicals that kill host cells. An example is the canker-causing fungus *Fusicoccum amygdali,* which produces fusicoccin, a toxin that causes wilting and cell death in many plant species. Biotrophs cause minimal cellular damage and require living host tissue. Hemibiotrophs, which initially keep host tissue alive and then change to the necrotrophic mode, are among the world's most devastating plant disease organisms. Potato blight, caused by the hemibiotrophic oomycete *Phytophthora,* was responsible for the Irish famine of 1846 and 1847, which resulted in a million deaths through starvation and emigration of another million or more people to the United States and other countries.[147] The wrong sort of senescence and death at the wrong time is a misfortune for an individual plant or plant community; but it can be catastrophic for crops and the humans that depend on them.

The LORD shall smite thee with a consumption, and with a fever, and with an inflammation, and with an extreme burning, and with the sword, and with blasting, and with mildew; and they shall pursue thee until thou perish.
Deuteronomy 28:22 *King James Bible*

Powdery and downy mildews are specialised biotrophic fungi that attack many crop species and other plants. They draw nourishment from host cells, which must be viable, often through a feeding structure called a haustorium. One way in which plants react to this kind of biotrophic attack is to initiate senescence, thereby denying the pathogen access to the products of photosynthesis. The pathogen commonly responds by secreting cytokinins, anti-senescence hormones that prevent yellowing and sustain the host cell's capacity to fix carbon dioxide. The leaf becomes a battlefield, the scene of an arms race with senescence-controlling hormones as the weapons. Delayed senescence in a zone surrounding the infection site takes the form of a 'green island', a group of zombified photosynthetic cells that retain chlorophyll against a background of yellowing leaf tissue. Sometimes there are beautiful patterns of concentric yellow and green areas where waves of hormonal attack and defence have ebbed and flowed. The bond between mildew and senescence is a metaphor for corruption and death with biblical resonance. Hamlet berates his mother for marrying the fratricidal Claudius: 'Here is your husband, like a mildewed ear/Blasting[148] his wholesome brother'.

Leaf senescence is a vulnerable stage in the plant life cycle. Resources recycled from older tissues to new young organs such as developing seeds or root tubers, could be intercepted and absorbed by a biotrophic pathogen. In contrast to green island organisms, some pathogens, such as the highly host-specific bacteria of the genus *Pseudomonas*, produce toxins that stimulate chlorosis, a senescence-like yellowing of host leaf tissues.[149] Viruses that cause plant disease are also generally biotrophs. They are often transmitted from host to host via an insect or other invertebrate vector and, after invasion, they replicate in host cells and spread through the plant's system of interconnecting cell-cell bridges (plasmodesmata).[150] Symptoms of viral infection are chlorosis or necrosis (browning), mosaic patterns (chlorotic mottling) and stunting in the host. Because older tissues offer a feast to potential pathogens, plants commonly step up their defence pathways as they initiate normal non-pathological senescence. Endogenous plant hormones that regulate senescence are generally also mediators of resistance against disease organisms. It seems that host and pathogen are driven to pull every trick in the book as they struggle for supremacy and survival.

Have you thanked a crop scientist today?

We should take a moment to acknowledge some unsung heroes. Have you thanked a plant pathologist today? Or a plant breeder? You should, because they are the largely unnoticed people standing between us and hunger. Here's a story. Ug99 is a new, extremely virulent, race of the biotrophic stem rust fungus that was discovered in Uganda in 1999. More than 90% of modern wheat varieties are susceptible to Ug99. None of the rust resistance genes available at the time was effective against Ug99. It took about 10 years to find sources of partial resistance to Ug99 in the global wheat gene pool and to begin to combine them to produce an acceptable degree of resistance that would withstand rapid evolution of new races of virulent rust. We are familiar with the annual cycle of influenza vaccination, made necessary by the relentless appearance of new strains of flu virus arising from mutations in the global virus gene pool. Crop pathologists and breeders are engaged in a similar, though largely unappreciated, arms race with the pests and diseases that threaten our food.[151] The time between introducing a crop variety with resistance to an existing race of pathogen and mutation of that pathogen to overcome the host's resistance can be as short as three years. At any given time, the world may be at best just a decade away from widespread famine. Ug99 and a whole slew of diseases and pests are still out there, waiting for us to drop our guard.

Plants have the capacity to arm themselves to the teeth with antimicrobial biochemicals.[152] Such constitutive resistance to pathogens is generally weak in crop species. This is because the toxicity of antimicrobials often extends to humans, with the consequence that the potential to biosynthesise and accumulate them has been progressively bred out of food plants. This improves nutritional quality but at the cost of increased susceptibility to diseases – hence the need for agrochemical pesticides to substitute for the missing endogenous defence compounds. Think about this next time you hear someone extoll the virtues of 'natural food'. As well as pre-formed, anticipatory defences, plants call upon a range of physiological measures that are induced when infection threatens. The hypersensitive response (HR) is one of these. Elvin C Stakman[153] introduced the concept of hypersensitivity (or 'hypersensitiveness', as he called it) in 1915. In the HR strategy the host, through induced suicide of infected cells, neutralises invasion by pathogenic fungi and viruses. HR has the appearance of cell death and necrotic flecking. Sacrificing a cell, tissue or organ in the interests of survival of the whole organism is a common habit of plants and is an example of the Samurai Law – it is better to be dead than wrong. In its simplest form, the HR mechanism is a rather beautiful lock-and-key-type interaction between resistance (R) genes in the host and avirulence genes in the pathogen.[154] R genes have been targeted by crop breeders, who have found them to be important sources of pathogen resistance. The detailed physiology of the HR is complex, but two environmental factors critical for HR to work are directly implicated in the story of senescence: oxygen and light.

Chapter 11 Oxygen: can't live with it, can't live without it

Leaves make ROS. Oxygen atoms are blue, the hydrogens of hydrogen peroxide are yellow and the extra electron conferring reactivity and a negative charge to the superoxide anion is red[155]

Originally Earth's early atmosphere contained no oxygen. It was not until photosynthetic micro-organisms (cyanobacteria) evolved that oxygen production added appreciable quantities, probably well before 2.5 billion years ago.

David Beerling[156]

You are standing on a headland, looking at the sea. The chalk beneath your feet is oxidised carbon. The sand on the shore is oxidised silicon. The waves that break on the beach are oxidised hydrogen. The rusting hull of the trawler on the horizon is oxidised iron. Its funnel emits oxidised sulphur and nitrogen. High above your head, the rays of the sun are filtered out by the ozone layer, oxidised oxygen. You're smiling at this pleasant scene, revealing the oxidised phosphorus of your teeth. Oxygen is everywhere, and is a most needy element: the way its electrons are arranged means that the oxygen atom hates to be alone and will attach itself to almost any other element, including other oxygen atoms – which is why the air we breathe contains not O but O_2. All the O_2 in the atmosphere comes from the photosynthetic splitting of water molecules, H_2O, by algae and terrestrial plants. Oxygen's neediness makes it reactive. The various forms of elemental oxygen and its derivatives are collectively called reactive oxygen species (ROS). Some of the most intensively researched ideas about the mechanisms of biological ageing and senescence are centred on the regulatory or destructive properties of ROS.

Among the notable things about fire is that it also requires oxygen to burn - exactly like its enemy, life. Thereby are life and flames so often compared.
   Otto Weininger[157]

Reactive oxygen species, ROS, are normal intermediates in biochemical reactions essential for living cells, but they need to be kept under tight control if they are not to cause serious damage. The commonest forms of ROS are superoxide anion (formed in enzyme reactions occurring naturally throughout the cell); hydrogen peroxide (consumed and produced by many biochemical reactions, and an important player in cell defences against pathogen attack[158]); and singlet oxygen (produced by photosynthesising chloroplasts, as well as by exposing O_2 to ultraviolet light). An early event in the hypersensitive response to pathogen attack is production, by an enzyme at the cell surface, of superoxide which is a signal to activate host defences. Green tissues use light as the principal source of energy and emit oxygen as a product of photosynthetic carbon reduction. In senescence, photosynthetic energy is increasingly replaced by energy from respiration, an oxidative process.[159] The general trend towards increasing oxidation with tissue age is associated with a progressive buildup of ROS. Some authorities consider ROS accumulation to be causative; others think it is more a symptom of senescence or declining viability.

Reactive oxygen species (ROS) react with other molecules in living cells to propagate a cascade of harmful free radicals that can end up by wrecking or even killing the cell.[160] Plant cells are equipped with many levels of defence against damage by ROS and free radicals. Carotenoids, ascorbic acid (vitamin C) and glutathione are typical antioxidant compounds found in plant cells. Certain enzymes[161] are also deployed to mop up excess, potentially harmful ROS. Some models of senescence regulation propose that a programmed decline or reorganization of antioxidant capacity results in elevated ROS and a consequent increase in free radicals and pathological physiology. These hypotheses regard senescence as essentially a speeded-up version of ageing. ROS themselves may be signals that modify patterns of gene expression. For example hydrogen peroxide is a component of the regulatory systems controlling synthesis of the enzymes of its own metabolism. ROS are also an important point of contact between senescence and reactions to attack by disease organisms.[162] There are many distinguished biologists (not to mention cosmetics companies) who will tell you that stress responses, senescence and ageing can be accounted for almost exclusively in terms of ROS and free radicals. I'm not among their number, remaining unconvinced that the molecular mechanisms implicit in these hypotheses are truly physiological and causative rather than post-mortem and consequential.

I have no truck with lettuce, cabbage, and similar chlorophyll.

S.J. Perelman[163]

The hypersensitive response to pathogen attack is a kind of pseudosenescence.[164] Cells commit suicide under the influence of oxygen and light. For light to induce biological change it has to be absorbed by a photoreceptor, a pigment that collects photon energy and transduces it into chemical action. Tissues that are harmed by light are said to have been photosensitised, and the pigments responsible are phototoxic. An obvious point of contact between the photosensitivity of pseudosenescence and the yellowing that occurs in 'true' senescence is chlorophyll, the light receptor for photosynthesis. Chlorophyll in the wrong place can be phototoxic. Normally it is packaged harmlessly, tucked away within the chloroplasts of photosynthetic cells and surrounded by antioxidants and other restraints. But senescence requires chlorophyll's shackles to be removed so that the enzymes that dispose of it can get access. This is a delicate, and potentially dangerous, operation. Excitation by light of the free pigment and its coloured breakdown products leads to the formation of reactive oxygen species, photobleaching, pseudosenescence and cell suicide. Some forms of hypersensitive response capitalise on the phototoxicity of chlorophyll and its derivatives in order to cauterise the point of pathogen invasion.[165] There are herbicides that work in a similar way, exploiting chlorophyll's photosensitising properties. Chlorophyll derivatives can also photosensitise animals, including people.[166] Dietary chlorophyll getting into the skin of an albino, or an animal with a malfunctioning liver, can cause unpleasant skin lesions. Occasionally molluscs that feed on plankton can accumulate high levels of phototoxic chlorophyll derivatives. In Japan, it is said that excessive consumption of abalone may cause your ears to fall off. The reason is that these pigments build up in the skin and the light striking the top of the head can cause such severe lesions that the junction with the ear is eaten away.

A zebra does not change its spots.
　　Al Gore

If a plant has flecks, spots or blemishes, it's reasonable to conclude that it has been the subject of pathogen attack. But there are strains of some plant species that spontaneously develop localised areas of cell (pseudo)senescence, damage or death, even without a pathogenic biotic encounter. So-called lesion mimic mutants[167] are of great interest because they allow the genetic control and biochemistry of resistance processes such as the hypersensitive response to be analysed. More than thirty lesion mimic mutants of maize are known, and many more than that in the model species Arabidopsis. A practically useful characteristic of some lesion mutants is that, by virtue of the fact that their hypersensitivity machinery is partially or completely in the *always-on* state, they have a high degree of intrinsic pathogen resistance. There are advantages to be gained by breeding such resistance into the backgrounds of elite crop varieties, provided the presence of lesions does not significantly interfere with overall photosynthetic productivity. An example is the Sekiguchi lesion (*sl*) mutation in rice, which confers resistance against the devastating fungal disease rice blast. Another is the *mlo* (*mildew resistance locus o*) mutation of barley which, despite the necrotic leaf flecking it causes, is widely used in spring barley cultivars because it confers increased resistance against powdery mildew.[168]

As if the attentions of pathogenic fungi, viruses and bacteria weren't vexatious enough, plants also have to contend with animals that chew them, pierce them, trample them, mimic them, hide in them and lay eggs on them. In some ecosystems the fate of as much as 70% of plant biomass is to pass through the guts of animals. Constant attack from invertebrates results in physical and physiological symptoms that include senescence. Chewing insects can be responsible for spectacular plant tissue damage and death. Plagues of locusts can defoliate an entire crop covering several hectares within a day or so.[169] Tissue damage caused by chewing insects frequently permits secondary infection by necrotrophs. Aphids or thrips (in Simon Leather's words 'beautiful creatures – sap-sucking insects') rarely cause extensive physical injury, although heavy infestations can result in chronic shortages of the products of photosynthesis and severe reductions in growth.[170] While feeding, many sap-sucking insect pests also transmit viruses. Parasitic nematodes are among the most damaging invertebrate herbivores.[171] They infect and disrupt the architecture of root systems, and often cause large changes in the physiology of the whole plant, including altered patterns of senescence. Adaptations designed to deter tissue damage by grazing or browsing animals include prickles, spines and hairs, some of which are the developmental outcome of cell-specific lysigenous senescence. Damage caused by herbivores and parasites triggers local and systemic wound responses in plant tissues, and is often accompanied by the production of mobile or volatile alarm signal molecules.[172] Truly, as Darwin remarked, 'it is interesting to contemplate a tangled bank', the scene of 'the war of nature, from famine and death'. It really is a battlefield out there.

Pasture grasses are particularly well adapted for defoliation stress. Their shoot meristems are located low down, near the soil surface[173], and they have efficient senescence mechanisms for recycling nutrients to support recovery growth. In turn, co-evolution with plants adapted to defoliation and trampling has influenced the features that fit the grazers for a diet of vegetation – for example, teeth designed for biting and chewing foliage.[174] Herbivores must have digestive systems capable of processing high-bulk, low-nutrient content feed. The major grazing animals of agriculture, and the most abundant wild herd animals of natural grasslands and prairies, are ruminants – cattle, sheep, bison, wildebeest and so forth. Their defining digestive organ, the rumen, is a remarkable product of evolution, an anaerobic fermenter, located between mouth and true stomach, that turns plants into meat, milk, hide and fibre. Everything about rumen function is large-scale. The volume of a cow's rumen is up to one hundred and fifty litres. Along with the herbage the animal ingests, a similar volume of saliva is delivered to the rumen every day. Up to fifty litres of gas (mostly methane) is belched daily. Every millilitre of digestive liquor contains 5 billion bacteria, half a million protozoa and 50 thousand fungi. One minute a grass leaf is growing cheerfully in the sunshine, blowing in the wind at 15°. Next minute, along comes a cow and the leaf is plunged into anaerobic darkness at 39° and attacked by a cocktail of fermentative microbes. What happens next is an unexpected chapter in the senescence story.[175]

Forage's Adventures in Rumenland[176]

The microorganisms that live in the rumen of a cow or sheep need energy and will go to extreme lengths to get it. Ideally microbial energy requirements should be satisfied by sugars and lipids in the diet. But there's another potential energy source. Leaves contain protein – the protein that ultimately builds the proteins of the animal's own tissues or milk. It's preferable that leaf protein gets through the rumen intact and reaches the true stomach and intestines, where it can be efficiently broken down into amino acids, absorbed and converted into animal protein. But it frequently happens that the shock of being grazed and plunged into the stressful environment of the rumen is enough to trigger senescence-like reactions in leaf cells, including activation of their own protein-degrading enzyme systems. Faced with the choice between getting energy from relatively inaccessible cell walls and membranes on the one hand, or the products of protein breakdown on the other, rumen microbes will preferentially ferment the plant's free amino acids. When amino acids are fermented, the nitrogen they contain is converted to ammonia. Ammonia is poorly utilised by ruminants and emerges in the form of urine and malodorous high-nitrogen slurries and emissions, with dangerous consequences for air, soil and water quality. There's a more sinister side to this as well. Nitrogen that could and should have been captured as meat or milk is being lost from the animal, diminishing economic efficiency. In intensive livestock systems the remedy may be to supplement forage with extra protein; but if the additional protein is of animal origin…well, the BSE story is too well known to require further telling here. All this can be traced back to the cleavage of a single chemical bond, the one between the carbon and nitrogen atoms of amino acid molecules. On a global scale, everywhere you look, the making and breaking of carbon-nitrogen associations seems to mean environmental and economic trouble. It may be a bit of an exaggeration to reduce huge complex issues like BSE, agricultural emissions and intensification to the breaking of one 'bad' chemical bond; but as someone once said, inside every big intractable problem is a small approachable problem trying to get out.[177]

Chapter 12 Traffic signals and condemned cells

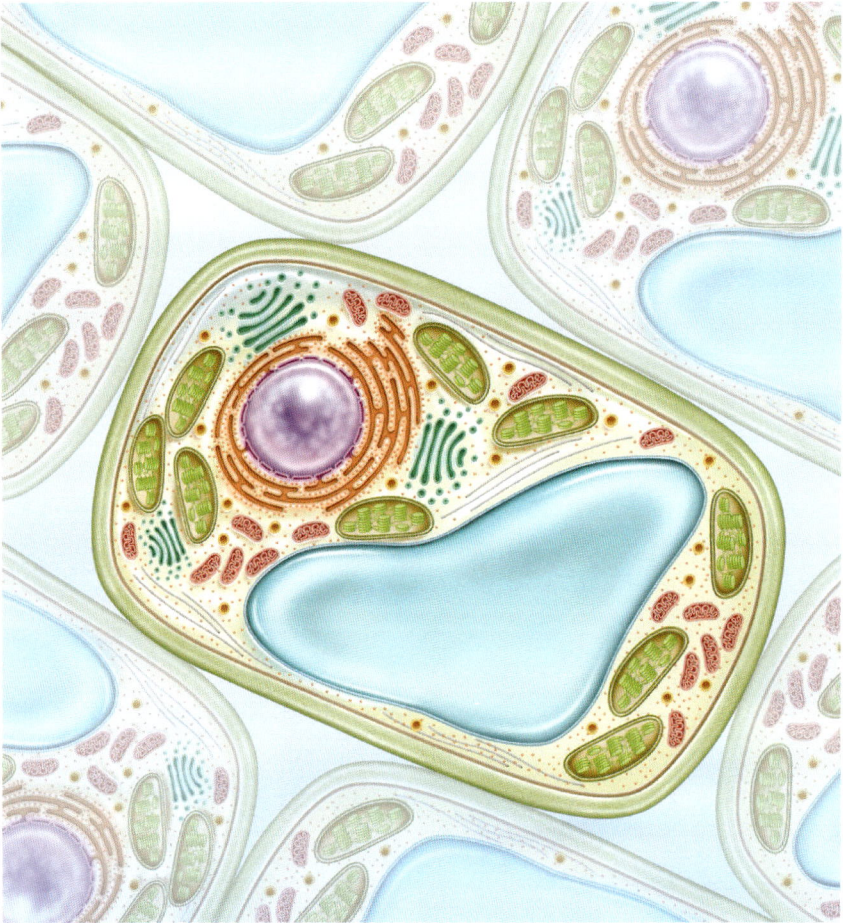

Green plant cell[178]

I have listed 16 different hypotheses that have been put forward to answer the question 'What is the adaptive advantage of leaf colour change in autumn?'

Marco Archetti[179]

Why do leaves turn red in the fall? The red pigments are anthocyanins, members of the same family of red, blue and purple phenolics as the biochemicals responsible for the colours of grapes, cherries, lollo rosso lettuce and violet petals. Senescence researchers have long brooded on the evolutionary origin and significance of red anthocyanin coloration in autumn leaves.[180] Demographic modelling, and some experimental observations, support the idea of coevolution with herbivores, in which foliar anthocyanins act as deterrent signals to insects and other predators. According to this view, the driving force for the development of colourful senescing foliage is similar to the coevolutionary visual interactions with pollinators and seed dispersers that led to the formation of pigmented flowers and fruits. Alternatively, or additionally, anthocyanin accumulation may be a plant response to, or insurance against the effects of, abiotic stresses. In particular, the hazards of excessive light, which can invoke pseudosenescence in the wrong place at the wrong time, can be quelled by employing anthocyanins as sunblockers. Among the biochemical intermediates in the pathway of chlorophyll breakdown is a phototoxic red compound[181] that never normally accumulates to more than negligible amounts, but which can build up in stressful conditions. It's conceivable that anthocyanins are an insurance against photosensitisation by this intermediate. In the words of Timbuk3 'the future's so bright, I gotta wear shades'. Autumn leaves are aesthetically pleasing and surely must mean something. But maybe not – after all, the yolk of an egg is a beautiful golden colour, but was never meant to be seen by anyone. Colour in nature is like this, only meaningful because we have eyes to see and minds and emotions to impart meaning. Perhaps autumn leaves are simply an example of what AL Kennedy has called 'unnecessary beauty'.

Chlorophyll, the portal through which photon energy enters the biological world, stands at the centre of perceptions of senescence. Its disappearance is the visual signature of a plant that is moving on to a new post-photosynthetic phase of development - senescence. The French chemists Pierre-Joseph Pelletier and Joseph-Bienaimé Caventou were the first to isolate the green pigment from leaves, in 1817. They called it 'chlorophyle'. John Ruskin hated the name. 'I wish they would use English instead of Greek words,' he grumbled in 1889. 'When I want to know why a leaf is green, they tell me it is coloured by 'chlorophyll', which at first sounds very instructive; but if they would only say plainly that a leaf is coloured green by a thing which is called 'green leaf', we should see more precisely how far we had got.'[182] In fairness to Ruskin, it indeed took quite a long time to get further. Not until the 1940s did Sam Granick and other researchers began to work out how plants make chlorophyll, building on earlier work on the biosynthesis of haem, the pigment of red blood cells to which chlorophyll is chemically related.[183] Robert Woodward completed the chemical synthesis of chlorophyll a in 1960 (and was awarded the Nobel Prize for this immense achievement in 1965).[184] By the 1980s we had detailed understanding of how chlorophyll is made in plant cells; but what about the pathways by which chlorophyll was dismantled? At that stage virtually nothing was known, in spite of the scale of seasonal senescence as a global event.

Colorless green ideas sleep furiously
Noam Chomsky[185]

The chlorophyll molecule is squarish and flat. Because of its atomic bonding structure, its electrons exist as a delocalised cloud above and below the plane of the molecule. Chlorophyll molecules form stacks (often called antennae) and their delocalised electrons become fused into, effectively, one big shared cloud. This means that when a photon interacts with any electron in the cloud, its energy is rapidly transferred through the whole antenna array until it reaches the photosynthetic reaction centres where it can be used to release oxygen from water and fix carbon dioxide. If for any reason photon energy can't be delivered to a reaction centre, light-excited chlorophyll becomes hazardous, a source of reactive chemical species that can damage or kill the cell. Before chlorophyll biochemistry in senescence was actively researched, it was assumed that the pigment disappeared by spontaneously bleaching as defences against wandering overexcited electrons became less effective with age. Now it's known that the removal of chlorophyll during true (as opposed to pseudo-) senescence is far from a passive letting-go process. When leaves turn yellow, their chlorophyll is being converted into colourless, photochemically inert breakdown products by an active, high-precision biochemical mechanism nicely matched to the risky job it has to undertake.

Catabolism, the converse of synthesis, refers to the unmaking of biomolecules. The products of catabolism are called catabolites.[186] The biochemical pathway and subcellular organisation of chlorophyll catabolism was largely worked out in the laboratory of Philippe Matile[187] in Zürich. I was fortunate to have been around when my late friend and his colleagues were patiently revealing what turned out, surprisingly, to be a detoxification mechanism – or, one might say, a kind of molecular-scale exercise in bomb disposal. One of the earliest advances was the isolation, from senescent barley leaves, of the first colourless chlorophyll catabolite. Its chemical structure showed that one of the steps leading to its formation must have involved opening up the intact chlorophyll molecule by cutting a particular chemical bond.[188] Furthermore, the fact that the breakdown product is colourless says that the delocalised electron system, which accounts for the pigment's green colour, must have been disrupted in the course of catabolism. We now know that these critical initial events in the delicate deconstruction of the chlorophyll molecule, rendering it photodynamically harmless, are carried out by a complex of four or five enzymes acting as a molecular machine – a sort of bomb-defusing nanorobot. These reactions occur inside the chloroplast, the subcellular structure within which photosynthesis happens. The initial colourless product of the chlorophyll breakdown machine is shipped out of the chloroplast into the central vacuole, the cell's dumping ground and storage container, undergoing further chemical modifications on the way. The catabolite first isolated from barley is one of the terminal products that accumulate in the vacuole. Subsequent research on the fate of chlorophyll in senescence is discovering an unexpectedly rich variety of chemical structures and fates of catabolic derivatives in different plant species.[189]

If we look inside the senescing cell, we find a busy place with information and material passing back and forth between different structures. A cell is made up of a number of compartments – rather like a railway carriage of former times, the classic crime scene in the golden age of mystery fiction.[190] Cell compartments (organelles is the technical term) are the settings for the biochemical processes that lead to senescence, terminal differentiation and death. Organelles are embedded in the cytosol, the gel-like background material of the cell. The single nucleus is the cell's control centre, where genes reside. Mitochondria are the numerous small organelles that provide energy. Plastids are unique to plant cells and come in a range of variants. Of particular significance in senescence are the pigmented plastids: chloroplasts (green organelles where photosynthesis takes place) and chromoplasts (the yellow, orange or red plastids of flowers and fruits).[191] A further distinctive feature of the plant cell is a large water-filled central organelle, the vacuole, which carries out a range of essential jobs. It is responsible for hydrostatic control (wilting is the symptom of this function under stress). The vacuole is the site of storage and turnover for dissolved sugars, ions and waste products. It is also a highly lytic compartment, containing a cocktail of aggressive enzymes that break down biomolecules. Senescence is a time in the life of the cell when there are particularly active exchanges between cell organelles. This is visible at every scale, from the microscopic to the landscape. The yellowing of foliage is the signature of events in chloroplasts, as chlorophyll is rendered colourless and photodynamically inert. Leaves turn red when anthocyanin pigments accumulate in vacuoles. As it approaches the terminal phase of its life, the green plant cell certainly does not go gentle into that good night.[192]

Your molecular structure is really something fine:

A first-rate example of functional design.

Mose Allison[193]

Chlorophyll has a job to do, but cannot do it (indeed, it is photodynamically lethal) if it simply sloshes around freely in the cell in a disorganised way. Accordingly, the pigment is normally structured into light-collecting antennae and photosynthetic reaction centres by association with proteins in the internal membranes[194] of the chloroplast. Such assemblages of chlorophylls and proteins not only precisely configure light harvesting and the conversion of photon energy to bioenergy, they keep the pigment's phototoxic tendencies under control.[195] Collectively, the chlorophyll-binding proteins account for quite a high proportion – maybe even a quarter - of the total protein of green cells. This is significant because senescence is the period when the protein of older cells is broken down into amino acids[196], the mobile form of organic nitrogen in the plant's internal nutrient economy, and recycled by transfer to younger, growing tissues. Chlorophyll-binding proteins are a prime source of salvageable nitrogen. But simply sending in the wrecking crew to take these proteins apart would risk releasing chlorophyll from its constraints to unleash reactive oxygen and free radical havoc in the cell. The advantage to the plant of investing resources in building an intricate molecular machine for delicately breaking down chlorophyll becomes clear: it allows pigment-protein complexes to be unpicked safely and amino acids to be recovered and reused. Chlorophyll itself contains nitrogen – four atoms per molecule[197] – but on the whole plants do not seem particularly concerned about reusing it. If the terminal products of chlorophyll catabolism end up in the vacuole and they and their nitrogen are lost when the leaf is shed or decays, this may simply be the price the plant is prepared to pay for access to the much larger store of mobilisable nitrogen in the pigment-protein complexes. The closer one looks, the more subtle and finely tuned the interplay between different structures within the senescing cell appear to be.

Rubisco[198] sounds like a company that manufactures breakfast cereal. In a way, that's just what it is, because it's the informal name of the enzyme that fixes carbon dioxide during photosynthesis, and therefore it's ultimately responsible for breakfast cereals (and all the other food we eat). Chlorophyll and rubisco are the gateways through which solar energy and carbon enter the biosphere. Their disposal is also the central feature of plant senescence. Rubisco is the most abundant protein in plants – by some calculations, possibly the most abundant protein in the world.[199] It accounts for more than 60% of total leaf protein. The proteins with which chlorophyll is complexed also make up a large proportion, as much as a quarter, of leaf protein. Like chlorophyll and its associated proteins, rubisco is located in the chloroplast.[200] Senescence is the phase of development in which the proteins and other materials of older cells are taken apart and the breakdown products salvaged and redirected to build new young tissues. Chloroplasts (and, by extension, green cells, and leaves, and whole canopies) are therefore bifunctional. They start by being sources of fixed carbon during the photosynthetic phase of their lifecycles. Then, with the onset of senescence, they switch to a recycling function, supplying organic nitrogen and other mobilised resources for use elsewhere in the plant. We have learned a fair amount about what happens to chlorophyll during senescence; but despite the scale of the process, working out how plants dismantle and redistribute rubisco, pigment complexes and other chloroplast proteins is proving to be tricky.

I say to the Occupy protesters - you're occupying the wrong place
Sarah Palin

Chloroplast proteins are taken to pieces during senescence and the products of their catabolism are redistributed to new, growing tissues. At the same time, enzymes that break down proteins – proteases – become more active. The control of protein recycling in senescence, then, looks straightforward. Switch on proteases and watch chloroplast proteins begin into disappear. But it isn't so simple. Chloroplast proteins are in – yes – the chloroplast. But most of the proteases, including those that are activated in senescence, are in the cell vacuole. How do the vacuolar proteases and their presumptive chloroplast-located substrates get together? The answer is – nobody's really sure. At one time it was thought that the cell vacuole might engulf whole chloroplasts during senescence. Another proposal had the vacuole membrane rupturing, flooding the cell with lytic enzymes. Yet another idea was that protein breakdown in senescence is nothing to do with the actions of vacuolar enzymes, but instead is catalysed by a system of completely different proteases located inside the chloroplast. Recently, advances in microscopy have enabled the dynamics of organelle interactions within cells to be observed in much more detail than ever before. These have revealed small vesicles containing proteins moving between chloroplasts and the vacuole.[201] Chloroplasts and vacuoles: working out the flow of traffic between these organelles is the key to understanding the how, and maybe even the why, of plant senescence.

At first glance, the central vacuole of the plant cell doesn't look very interesting – a bag of water acting like the hydrostatic equivalent of the pneumatic inner tube of a car or bicycle tyre. It was Philippe Matile[202] who played a leading role in showing the true versatility and physiological importance of this organelle. His 1975 monograph *The Lytic Compartment of Plant Cells* was a landmark in the science of senescence, complementing his later definitive contributions to chlorophyll catabolism. Animal cells were known to contain lysosomes, small vesicles rich in catabolic enzymes.[203] Matile and others showed that the central vacuole has some of the enzymic features of a 'plant lysosome'. It is also a repository of antibiotics and toxins, which makes the vacuole a potent defensive weapon against pathogen attack – in effect behaving as the plant's immune system. A fungus that poked its infection thread into the cell would get a nasty shock when it encountered the lytic contents of the vacuole. So too a sucking insect that inserted its stylet. In a way, the vacuole is reminiscent of the Bocca della Verità, a round medallion of Pavonazzetto marble, about two thousand two hundred years old, located in the portico of the Roman church of Santa Maria in Cosmedin[204]. Legend has it that if you tell a lie while your hand is in the mouth of the sculpture, your hand will be bitten off. A pathogen, pest or parasite invading the plant cell vacuole with malign intent takes the Bocca della Verità test and risks similarly harmful consequences. The lytic and phytotoxic nature of the vacuole is protective against invasive threats from outside; but for the cell that contains it, the organelle is nothing less than a potential suicide pill.

this dissolution of self, this autophagy
Alaya Dawn Johnson[205]

Physical damage can kill a cell, as can chemical trauma, for example in the form of the toxins secreted by a necrotrophic pathogen. Cells can also control the course and outcome of the terminal phase of their lifespan through an act of immolation initiated and executed from within. Cell senescence and death are essential for normal development and for adaptive or acclimatory responses to environmental stress. Most of the modes of senescence discussed in this book are variations on the general theme of cell self-determination: leaf yellowing, fruit ripening, lysigeny, development of wood, cork, spines, pollen and aerenchyma, formation of glands and ducts, the hypersensitive response, root turnover. Autophagy (literally 'eating oneself') is a particular form of termination behaviour of cells found across the whole range of organisms, from animals to fungi to plants[206], and one that is turning out to be of special significance for aspects of ageing. It is controlled by a network of molecular switches responsive to developmental signals and environmental stresses. During autophagy cell contents, including intact organelles, are engulfed by vesicles called autophagosomes. In plants, autophagosomes are taken up by the cell's large central vacuole. In animals they fuse with lysosomes. The contents released from autophagosomes are broken down by the lytic enzymes of the vacuole or lysosome that engulfed them. More and more cases of plant autophagy are coming to light. It has been observed in the short-lived flowers of Japanese morning glory, during the hypersensitive response to pathogen attack, in nutrient-starved cell cultures, in tissues of germinating seeds, and during the terminal differentiation of woody tissue. More research is needed but vesicles, autophagic or autophagic-like, look to be the best bet for getting proteins from chloroplasts and other parts of the senescing cell to the lytic environment of the vacuole for salvage and recycling.[207]

Chloroplasts (and vacuoles) are the sites of much of the action in senescing cells. The replacement of greenness with yellows and reds is the visible symptom not only of the catabolism of chlorophyll but also the stripping down and reallocation of chloroplast proteins, notably rubisco and pigment-binding complexes. The bifunctional nature of chloroplasts, as sites of photosynthesis and repositories of raw materials for recycling, extends beyond chlorophyll and protein. Each chloroplast has its own genetic system, comprising DNA (carrying a number of genes) and a protein-synthesising machinery based on RNA copies of this DNA. Nucleic acids have an almost sacred status in biology by virtue of the genetic information they encode. But they are also chemical in nature like all the other stuff that cells are made from. And just as senescence is the process by which protein stuff and pigment stuff are dismantled and shipped out for re-use elsewhere, so too does it subject chloroplast DNA and RNA to salvage and reallocation. Nucleic acids are rich sources of phosphorus which, with nitrogen, is a leading currency in the plant's internal nutrient economy. And now we come up against a similar enigma to that facing our attempts to understand protein mobilisation during senescence. The enzymes that take DNA and RNA apart are constituents of the lytic brew that lives outside the chloroplast in the vacuole or the membrane systems of the cytoplasm.[208] How do the nucleic acids of the chloroplast and the corresponding vacuolar catabolic enzymes meet? If I had to guess (and it has to be a guess because experimental evidence is scarce) I would predict it will turn out to be some variation of autophagy, in which vesicles carrying material from the senescing chloroplast are engulfed by the vacuole and deliver nucleic acids, together with proteins, lipids and other stuff, for catabolic processing.

Chapter 13 Staying green

Mendel[209]

Gregor Mendel (1822-1884), botanist, mathematician and physicist, of the Augustinian order at the Monastery of St. Thomas in Brno, is often called the Father of Genetics. There's an argument for identifying him as the Father of Plant Senescence too. Against his immense biology-wide achievement in creating the foundations of heredity, the physiological details of the studies on peas he carried out between 1856 and 1862 are often overlooked. Using genetics explicitly to understand plant senescence did not begin until over a century after Mendel's experiments in the monastery garden at Brno. It was as a young researcher in 1972 that I first encountered a botanical curio – a line of meadow fescue grass with leaves that appeared not to turn yellow at any time. Following the broad principles of Mendelian analysis, I set about crossing and back-crossing green and yellowing fescues. Simple though the experiments were, it was thrilling to be able to verify 3:1 and 1:2:1 ratios of yellow and green types in the progeny, just as Mendel did in his peas.[210] Mendel wasn't particularly interested in senescence, any more than he knew anything about genes and DNA. But among the seven traits he studied was colour of the mature pea seed – green or yellow. A pea comprises two hemispherical organs (the cotyledons) enclosing the embryonic axis. A cotyledon is developmentally a modified leaf – a fat leaf, you might say. As I counted green and yellow leaves in my fescue inheritance study, I was conscious that Mendel had done the same thing with his pea cotyledons more than a century earlier. He had identified for the first time a 'senescence gene'.

The abbot G. Mendel, a brilliant fellow,
Wondered why some peas were green and not yellow.
He knew nothing of genes, let alone DNA,
But he worked out the rules of the game anyway.

    Helen Ougham[211]

Gregor Mendel showed that the colour of a mature pea is determined by what we would now call a single Mendelian gene. Normally the action of this gene would result in the disappearance of chlorophyll as cotyledons grow to full size, resulting in a yellow pea. If the gene isn't working for some reason, chlorophyll remains and the pea is green. This is precisely what I found when I analysed the inheritance pattern of leaf senescence in normally yellowing and persistently green lines of meadow fescue grass. Senescence in the latter is defective because a gene normally responsible for yellowing is faulty.[212] This led to the notion that, if the 'green gene' in peas and in meadow fescue is functionally the same gene, doing a similar job in seeds and leaves, then the foliage of green-seeded peas might be expected to retain chlorophyll during senescence. We confirmed this to be the case on one of my visits to Zürich.[213] Some time during the 1960s (the earliest record I can find is in a 1962 publication from a Dutch agricultural station) plant breeders and crop physiologists started referring to a variety with persistent chlorophyll as 'stay-green'. As awareness of the trait grew, so too did the range of species and crop products displaying the character get broader. For example, the flageolet variety of the common bean *Phaseolus vulgaris* is a stay-green type, with seeds that are green at maturity. Most varieties of frozen peas these days are stay-green. Green flesh tomato and chlorophyll retainer bell pepper varieties, in which colour change in fruit ripening is blocked, are stay-green types. The list of stay-greens is growing steadily. But as physiologists and geneticists began looking more closely at the trait, it became clear that there is more than one way (and one gene) to stay green.[214]

As leaves turn yellow during senescence, chloroplasts are taken apart and the material reclaimed for use in young, growing parts of the plant. What happens to the chloroplasts of Mendel-type stay-greens like meadow fescue or peas?[215] It seems that they get partially dismantled. As in normally yellowing types, rubisco, many other chloroplast proteins, nucleic acids, carbohydrates and membrane lipids are catabolised, and photosynthetic capacity declines. But chlorophyll, being largely unscathed, has the effect of stabilising (and being stabilised by) the protein complexes it is associated with in the photosynthetic membranes of the chloroplast. Senescent stay-green leaves of this type superficially look like they should be photosynthetically productive, but are in fact physiological cadavers. They have been called 'cosmetic stay-greens'. In retrospect, this is a rather dismissive and misleading term (I write as someone responsible for introducing it), because they are interesting examples of how genes specify traits, and therefore are more than simply good-looking corpses. In most cosmetic stay-greens the trait can be attributed to a genetic mutation that blocks chlorophyll breakdown. Chlorophyll is catabolised by a multi-enzyme molecular machine that deftly extracts the pigment from its environment in the photosynthetic membrane of the chloroplast. The defective gene in Mendel's green pea and stay-green meadow fescue has been identified.[216] It's now called *SGR*. *SGR* encodes an essential protein component of the chlorophyll catabolism machine. If this protein is missing because *SGR* is faulty, as in green peas and stay-green fescue, the pigment breakdown complex cannot assemble and chlorophyll remains unbroken. We have good, and growing, understanding of the genetic and biochemical features of cosmetic stay-greens. But there is another set of stay-greens, much more diverse in nature, and of great importance in crops, that differs fundamentally from the cosmetic type: the functional stay-greens.

Cosmetic stay-greens retain some subcellular structure during senescence. Functionally, however, they go through the transition from source of photosynthetically fixed carbon to recycled nitrogen and other salvaged nutrients according to the same developmental schedule as normally yellowing types. What we can call, for brevity's sake, the C-N transition point is the initiation of senescence. A genetic variant in which this transition is delayed but the phase of yellowing and N mobilisation then proceeds as normal will be functionally stay-green. Another sort of functional stay-green is one where the C-N transition occurs at the usual time but subsequent yellowing and recycling is slower. Yet another can be both late and slow.[217] In the words of the great plant geneticist Barbara McClintock[218] 'it is a matter of timing the action of genes.' Because the duration of the photosynthetic phase in the life of the canopy is extended, the functional stay-green trait is one of the agronomically desirable attributes of most highly productive crop varieties. The cosmetic stay-green character is the expression of a faulty gene that encodes a step in the pathway of chlorophyll catabolism. Genetic control of the C-N transition can be more complex, involving genes that regulate the timing and rate of senescence and which act by turning other genes on and off. Identifying, isolating and determining the function of the different kinds of senescence-associated gene requires a combination of Mendelian methodology, genetic mapping, molecular analysis of gene expression and physiological measurements of senescence variants.[219]

If senescence and differentiation are aspects of the same process it is possible that there has evolved a group of genes which are repressed during development and normal functioning but which become active at a pre-determined time in the life of the organ or organism…

Harold Woolhouse[220]

Selective senescence of plant tissues, organs and individual cells is necessary for the normal development and survival of plants. Senescence and death of the whole individual is a normal part of the monocarpic lifecycle. The purposeful nature of the senescence that takes place during development, adaptation and propagation tells us it must be a tightly regulated process – that is, it must be programmed. In biology, where there's a program, there are genes. Some genes that participate in the senescence program are identifiable by classical Mendelian inheritance analysis. Cosmetic stay-greens are examples. But application of the tools of molecular biology, which take us directly to DNA, have revealed many more potential participants in the senescence program. Comparing the genes that are read out before the initiation of senescence with those that are active as the syndrome proceeds has revealed a large number (running into many hundreds) of senescence-associated genes (*SAGs*). Some genes are turned off (down-regulated) when senescence starts. Genes controlling chloroplast assembly are examples of down-regulated *SAGs*. Other *SAGs* are up-regulated, becoming more active as the senescence program is initiated and executed. One such is the cosmetic stay-green gene *SGR*, which encodes a critical component of the multi-enzyme complex that removes chlorophyll. Others include genes specifying degradative enzymes of the cell's lytic system for recycling proteins, nucleic acids and other salvaged materials, and genes determining defences against pests and pathogens. A class of *SAG* of particular interest for crop production and targeted plant breeding comprises the genes that regulate expression of other genes, thereby actuating or closing down entire physiological networks.[221] Variants of such genes give rise to functional stay-green traits, in which the senescence program starts late or runs slowly or both.

I went from adolescence to senility, trying to bypass maturity.
Tom Lehrer[222]

Senescence is programmed, and genes represent the code that makes things happen (or not happen) in the right places at the right times. But before the program can run, the cell, organ or organism must be ready. It must be competent to respond to the stimulus that says 'senesce now'. The progression from incompetence to competence is a developmental event marking the transition from juvenility to maturity.[223] A familiar example of juvenility is marcescence, where leaves that senesced in the autumn are not shed and stay attached to immature branches until displaced by the next generation of foliage. Many beech hedges are marcescent. Failure to shed senescence leaves is a juvenile character, seen in a number of tree species, including oaks and hornbeam as well as beech. The juvenile phase can be quite prolonged – up to sixty years in some oak species. Juvenility isn't confined to trees. Even the plant scientist's favourite model species, the small weed Arabidopsis, has to achieve a certain age before it can be induced to senesce. Some insights into how the juvenility-maturity switch is controlled have come from experiments on ivy. If tissue taken from a juvenile ivy plant is cultured, plants regenerated from it have stable juvenile characteristics (lobed leaves, high growth rate, red pigmentation, climbing habit, no flowers). Mature ivy tissue yields adult-type regenerants (unlobed leaves, low growth rate, no red colour, upright or horizontal growth, flowers), but can be induced to revert to juvenility if treated with the plant hormone gibberellin. Stable propagation of a reversible physiological state such as juvenility is a symptom of epigenetic control. Epigenetics, an area of intensive current research, is the term for heritable changes in gene expression (active versus inactive genes) without changes to the underlying DNA sequence.[224] Environmentally imposed epigenetic variation has echoes of the long-discredited theory of inheritance of acquired characters proposed by Jean-Baptiste Lamarck (1744-1829). As epigenetics has moved into the biological mainstream I have a mental picture of Lamarck, in some unregarded corner of the afterlife, raising a glass (perhaps in the company of Karl Marx) with the toast: 'I told you so'.

Chapter 14 Sources, sinks and calories

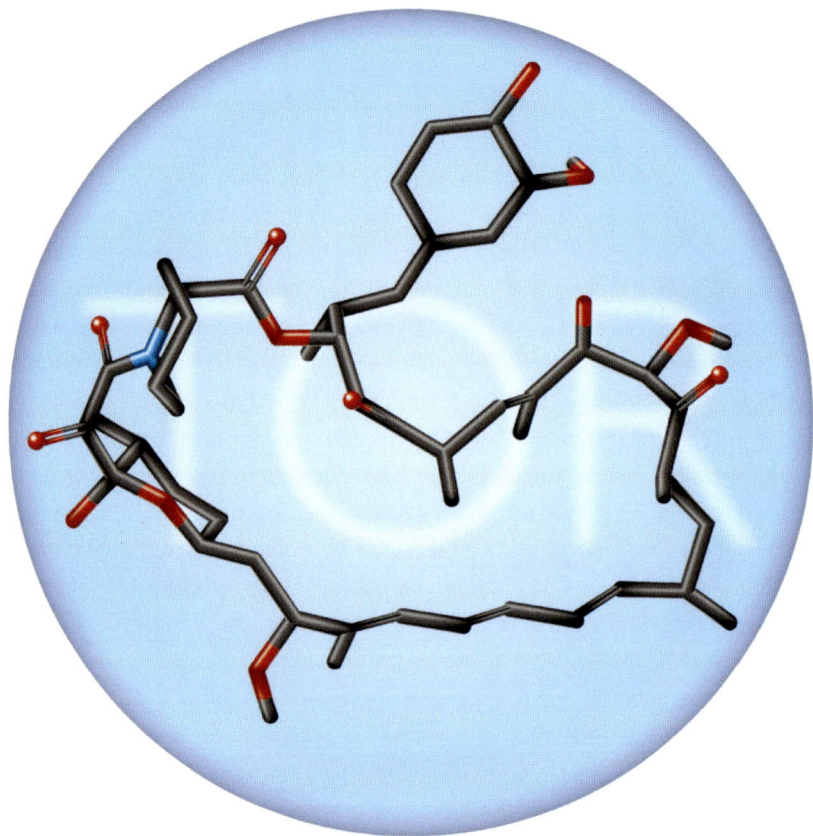

Rapamycin[225]

For it to grow, a developing seed or a young expanding leaf or a swelling root tuber must get nutrients from older parts of the plant. How do such sinks - destinations for materials transported from elsewhere – communicate their needs to sources of these materials? The question of source-sink relations is a central one in plant physiology. Before the onset of senescence, leaves are sources of carbon fixed by photosynthesis. During senescence, photosynthesis declines and the leaf becomes a source of nitrogen and other salvaged nutrients. How are supply and demand balanced? To put it vulgarly, does the source blow or the sink suck? The answer is that there is no single answer. It depends on the plant, its phase of development and the state of the environment. Where senescence is regulated by sink demand, then removing or inactivating sinks will halt senescence. Repeatedly taking the flower buds or pods off a soybean plant in the reproductive phase converts this monocarpic annual into a large, long-lived beanless beanstalk.[226] In the matter of source-sink regulation soybean, then, is a 'sucker'. Some species display a remarkable response of source leaf senescence to sink removal. Chopping off the shoot above the lowest, oldest leaves of a full-grown tobacco plant doesn't just halt senescence – it results in yellow leaves, even if they contain virtually no measurable chlorophyll, returning to full greenness. This reversibility of sink-induced senescence has been observed in a number of plant species. I suspect that, by employing the right experimental treatment, it should be possible to show regreening to be a near-universal phenomenon. Reversibility challenges the idea that senescence of green tissues is some kind of programmed death event. It certainly can, and usually does, result in death; but senescing tissue, far from dying, is completely viable and must retain physiological and genetic integrity to an advanced stage.

You do not often get the chance
Of seeing sugar brokers dance
WS Gilbert[227]

Senescence of source leaves can be stopped (soybeans) or even reversed (tobacco) by eliminating the influence of strong sinks. The source-sink relationship isn't always so clear and dramatic, however. Removing the developing cobs from maize, grains from wheat or fruits from bell peppers, if it has an effect on source foliage at all, tends to accelerate senescence. It might be possible to reconcile these contrasting source reactions to alterations in sink demand by considering how plant cells sense the availability of fixed carbon and nutrients. The products of photosynthetic carbon fixation are moved around the plant in the form of sucrose – the same sugar you stir into your tea and sprinkle on your cornflakes. There is growing evidence that sugars exert a regulatory influence over leaf senescence.[228] Variations in source-sink relationship between different plants may reflect a diversity of responses to sugar. Before it can be taken up on arrival at a sink, sucrose has to be cut in half by invertase, an enzyme that lives in the plant cell wall. The products of invertase action, glucose and fructose, enter into the cell and become part of the cell's biochemistry. Plant cells have separate molecular sensors for sucrose, glucose and fructose. These detect changes in the ratios of the different sugars and up- or-down-regulate the expression of gene networks accordingly, including those specifying senescence.[229] In the life of plants, sugar is truly multifunctional: material for building biomolecules, a source of bioenergy for physiological processes, and an information-broker, mediating in long- and short-range communication.

Plants, like animals, have hormones, but, as ever, plants do things their own way. Of the half a dozen or so chemical families of plant hormones, two or three have critical roles in the control of senescence.[230] The senescent lowest leaves of a tobacco plant will respond to cutting off the shoot above them by gradually recovering their green colour. They re-green much quicker if they are treated with plant hormones of the cytokinin family. The potency of cytokinins as anti-senescence hormones has been shown in a classic experiment carried out by Susheng Gan and Richard Amasino. They created a tobacco plant containing a gene that directs the synthesis of cytokinin. This gene was under the control of a genetic switch (promoter) that activated it only during senescence. The result was a frustrated plant – the more its leaves tried to senesce in response to source-sink or environmental signals, the more cytokinin they made and the more severely the symptoms of senescence were suppressed. Trapped in this autoregulated loop, the foliage of these plants remained green for a prolonged period, ultimately dying without ever senescing first.[231] Cytokinins are normally produced in roots and transported to leaves. Tissues with the highest cytokinin levels are the strongest sinks and attract the majority of nutrients by out-competing less active structures. In many plants at the fruiting stage of development, cytokinins from the root are redirected into developing seeds instead of into leaves. As the seeds become stronger sinks, nutrients are diverted to them and away from the leaves, which react by becoming senescent. This combination of hormonal and nutritional control is the basis of source-sink communication and the diverse patterns of foliar senescence in the various plant life-cycles.

Life-promoting TOR signalling seems also to contain seeds of death.
Mikhail V Blagosklonny & Michael N Hall[232]

Rapamycin is a drug of bacterial origin. As with many microbial bioactive compounds, it has a complicated barbed wire-like chemical structure. Rapamycin has been used in cancer treatments and to suppress immune rejection after transplant surgery. A property that got the gerontology community excited is that it can extend the lifespan of mice. Typically, at a time when survivorship of a population of untreated mice hits 0%, up to 40% of rapamycin-treated mice are still alive. One of the cellular proteins that rapamycin interacts with is called TOR – target of rapamycin. TOR is emerging as a point of convergence for a regulatory network coordinating energy status, sugar content, nitrogen availability, cell fate and longevity. The architecture of the TOR signalling system of animals and fungi is conserved, and that of plants will probably turn out to be similar too.[233] TOR is absolutely essential for development, controlling cell growth and proliferation. But once differentiation is complete TOR causes age-related deterioration, and lifespan can be increased by reducing TOR signalling. It looks increasingly like TOR is a nutrient-sensing regulator of plant growth and senescence too. One of the pathways through which it exerts its effects is autophagy, the organised self-digestive process that recycles obsolete cell structures and molecules. If TOR is confirmed as a central player in the control of nutrition, development and senescence, we may be edging closer to defining one of the universal mechanisms of biological ageing.

The nutritionist Clive M. McCay[234] showed that restricting the calorie content in the diet of lab rats could increase their lifespan. Similar observations have been reported for animals from a range of taxonomic groups. There is a lot of current research activity on caloric restriction, not least because of the headline-grabbing clinical and sociological problem of obesity. It is clear that excess calories are bad for health and life-expectancy. The contrary idea that restricting calorie intake can extend life is more controversial, however.[235] As for how diet could determine lifespan, attention is focused on resource availability in relation to age-related deterioration, where plants, fungi and animals may share functions and mechanisms. In particular, a picture is taking shape that links nutritional state, hormone action, growth and deteriorative ageing, facilitated by the physiological regulator TOR. Nutrient availability modulates the state of TOR, which in turn controls cellular repair and maintenance during development, and recycling during senescence, through variations in autophagy function. Excessive nutrient intake results in TOR-mediated suppression of autophagy, which in turn contributes both to obesity and to the accumulation of damaged and harmful cell constituents and structures. TOR also interacts with the insulin pathway: type II diabetes is another symptom associated with obesity and compromised lifespan.[236] Sorting the biological generalities from the organism-specific elements of these physiological and pathological processes will keep researchers busy for a while yet.

The combination of TOR, nutrition and hormones could be the biology-wide common factor in senescence and ageing. But before we get too excited, we need to remember the contrasting relationship of multicellular plants and animals to the capture and use of resource and energy. Some theories consider age-related physiological deterioration and mortality to be a reflection of a progressive imbalance in the relationship between maintenance and development.[237] TOR signalling is at the heart of this relationship. In general however, plants, unlike animals, are material- and energy-rich, capturing and deploying resources in a manner that has been described as promiscuous and even pathological.[238] Resource allocation between repair and growth in plants is therefore likely to mean something quite different in animals.

Chapter 15 Food for thought

Peppers[239]

By the end of this story, you won't believe how much you know about corn.
William Kamkwamba[240]

Ever since Thomas Malthus[241], even that part of the human race fortunate enough not to have to worry about where its next meal is coming from has realised that figures for global food supply and population don't add up. The world's population is increasing by about 80 million a year. At the same time, 10 million hectares of cropland worldwide are abandoned annually due to soil erosion and a further 10 million hectares are critically damaged each year by salinization. In 1960 each member of the world's population was supported by an average of 0.5 hectares of agricultural land; today the figure is 0.23 hectares and falling. That much is undeniable. And yet Malthus is out of fashion with policy-makers. At least in part this is because there have been repeated cries of 'Wolf!' in his name. Successive agricultural innovations, from Turnip Townsend and the four-year rotation system in the 18th century to the Green Revolutions of the 20th have confounded predictions of Malthusian disaster. But to be able to carry on doing more and more with less and less, every day around the world men and women are in fields, labouring to solve the unfavourable population-crop equation. To do this they need every tool that agricultural science can provide. As well as farmers, agronomists and researchers, our crops have to work harder. We need agricultural species to capture and convert resources ever more efficiently, offsetting the dwindling area of cultivatable land by extending the growing season and exploiting stressful environments. Crops must establish early and quickly, through rapid growth and development of the whole plant and its parts. They need to be adapted and adaptable to non-optimal environments. And they must be prolific, yielding usable products of high pre- and post-harvest quality. Senescence is an essential element in each of these aspects of crop performance.[242]

The improvement of yield potential in many crops has been associated with a greater duration of photosynthetic activity.

Lloyd T. Evans[243]

Modern agriculture has succeeded in confounding Malthus and feeding more people to a higher standard than ever before. But in doing so, it incurs economic, energetic and environmental costs that are almost certainly unsustainable. It therefore becomes all the more urgent to answer the question how has agriculture achieved its success-at-a-price? If there were a single answer (and of course there isn't) then it would be by keeping leaves alive longer. If you want to improve your crop, get its solar panels (foliage) out into the light as fast as possible and keep them there as long as you can. To understand the physiology of crop yield, it is essential to distinguish between durations and rates. As a general rule, the rate of photosynthesis (expressed on just about any basis you care to choose) notably falls short of predicting anything consistent about the useful output of most crops. The increased yields of modern 'improved' cultivars of wheat compared with the old 'inefficient' varieties they replaced have been achieved with no increase - even perhaps a slight *decrease* - in photosynthetic performance. On the other hand, the length of time the canopy can be retained in a green state is usually a good correlate of yield. In one analysis of field-grown maize under different fertiliser and irrigation treatments, it was calculated that each additional day of canopy green area duration could increase grain yield by around 5%. The equivalent figure for wheat is about 1% per day. Green area duration is one of the central concerns of crop agronomy and improvement. Once the canopy is established, its duration is fixed by the onset and progress of senescence.[244]

The productive life of the crop canopy has two phases. While green and photosynthetically active it is assimilating atmospheric CO_2 and making biomass. The carbon capture phase is followed by senescence, during which nitrogen and other nutrients are recycled. To control the transition between the two functional ages in the life of the canopy is a prime practical objective in agriculture. Could there be some kind of genetic 'master switch' that, once identified and analysed, will enable cessation of C acquisition and initiation of N salvage to be manipulated at will? This is the philosopher's stone of crop senescence research. Advances in molecular biology have brought us to the point where the powerful contemporary tools of genomics and systems biology can be used to define a global regulatory regime for senescence. In this way, we can recognise particular transcription factors (genes that turn on and off whole networks of subsidiary genes) that occupy strategic positions in pathways of developmental and environmental regulation.[245] Transcriptional regulators of the so-called NAC family are examples of particular interest for crop production.[246] They have given an insight into genetic changes associated with domestication of wheat, and direct evidence in realistic field conditions supports their role in altering senescence, yield and quality in this species. During the human-mediated evolution of wheat, genes of the NAC family became deactivated by selection for delayed senescence. As a consequence, photosynthetic duration and grain yield are greater in domesticated than in wild wheats. But there is a down-side. A prolonged carbon capture phase means delayed senescence and less effective salvage of nutrients. Compared with their wild relatives, the grains of domesticated wheats are highly floury (stuffed with starch, a direct product of more photosynthesis); but they are relatively impoverished in nitrogen, zinc and iron, nutrients that are recycled during senescence.[247] This trade-off between on the one hand, yield based on photosynthetically derived bulk carbon and, on the other, crop quality dependent on nutrients recycled during foliar senescence, presents agronomists and plant breeders with a nice conundrum.

Eventually most plant breeding may be based on ideotypes.
Colin M. Donald[248]

It's hazardous for a biologist to use the word 'design' because it has echoes of 'intelligent design' and it gets creationists over-excited. But it's allowable to talk of design if you're a plant breeder and you are thinking of the ideal set of traits for the crop plant you'd like to produce (I almost wrote 'create', but that might also get the evolution-deniers hot under the collar). Such a virtual plant, combining a set of optimal attributes, is an ideotype. What might the ideotype of a cereal species look like? For high yield it needs to photosynthesise over a prolonged period and transfer the maximal amount of fixed carbon to the developing grain, where it is stored as starch. This means establishing the canopy quickly. But because the canopy is there to intercept as many photons as possible, there shouldn't be so much foliage that leaves shade each other. In the crop physiologist's jargon we need an optimised canopy architecture: sufficient foliage to make a closed canopy and no more. It's also important that there's enough storage capacity in the developing grains to receive the fixed carbon produced by our highly productive canopy, otherwise sink-source signalling may slam the brakes on. Then, once the carbon capture phase of canopy development is completed, we need fast and efficient senescence so that there is maximal recovery of salvaged nutrients, contributing to good grain quality as well as yield. Add to these attributes the requirement for disease resistance, root systems with optimal water and nutrient use efficiency, and all-round stress tolerance and you present the breeder with a substantial challenge.[249] As Norman Borlaug, the father of the Green Revolution, said: 'There are no miracles in agricultural production.' Fortunately for us, breeders have been good at meeting the challenge. But in some crops, we are now so close to the ideotype that further improvement is getting increasingly difficult. Breeding for senescence-related traits has made an important contribution to the spectacular increases in yields of maize during the last century. It seems, however, that breeding for delayed senescence in maize, and probably in rice too, has gone about as far as it can for now, and current ideotypes focus more on sink capacity, plant architecture and resistance to pests, diseases and stress.[250] But improvement of other species, and in areas of crop systems other than primary production, continues to benefit from an understanding of senescence mechanisms and control.

It is sad to grow old but nice to ripen.
>Brigitte Bardot[251]

When we talk of grain ripening at harvest time and grapes ripening on the vine, we are recognising that ripening, a stage in development following morphological maturity, is one of a family of terminal senescence processes shared by green plant tissues. Physiological events in ripening fruit are variations on the theme of senescence, resulting in colour, texture, taste and odour changes that make fruits attractive for dispersers and for human consumption. A clear difference between ripening and senescence is the limited degree of salvage and export of materials from the cells of fruit tissues. But the chlorophyll of fruits is lost by the same biochemical pathway that operates in yellowing leaves. In the fruits of many species, chloroplasts differentiate into chromoplasts, plastids within which yellow and orange carotenoid pigments are unmasked, and new carotenoids are synthesised.[252] The red carotenoid of tomatoes and bell peppers is lycopene. The multicoloured packs of peppers in the greengrocery sections of supermarkets perfectly encapsulate the sequence of carotenoid and plastid transitions that occur during ripening. Another source of fruit colour is the pigmented water-soluble phenolics (the same family of pigments accumulated by autumn leaves), including the red and purple anthocyanins of grapes, cherries and strawberries. The fruits of some tomato hybrids have anthocyanin, which overlays the usual red of lycopene to give a purple colour (a similar result has been obtained using transgenic technology).[253] Anthocyanin-enriched tomatoes are claimed to have health benefits and enhanced shelf-life. Fleshy fruits get softer as they ripen, as the result of activation of enzymes that break down their cell walls. Fruits become sweet during ripening by importing sugars and degrading starch. They develop complex flavours and become more fragrant as they ripen, producing a large number of volatile organic compounds. It's estimated that about four hundred different volatiles contribute to the distinctive odour of tomato fruit. The genetic and hormonal regulation of fruit ripening has been intensively researched.[254] And the symbolism of fruit in world cultures resonates through the ages, from the Forbidden Fruit of Genesis to Abel Meeropol and Billie Holiday's 'Strange fruit hanging from the poplar trees'.

Ignorance is like a delicate exotic fruit; touch it and the bloom is gone.
  Oscar Wilde[255]

When it comes to the climacteric, our mystification would meet with Lady Bracknell's approval. Fruits are classified as climacteric (exhibiting a ripening-associated respiratory burst, often accompanying a spike in production of the gas ethylene) or non-climacteric.[256] Apples, bananas and tomatoes are climacteric; oranges, lemons, strawberries and grapes are non-climacteric. We don't know what function the respiratory climacteric fulfils. For sure it generates bioenergy and is a source of organic molecules for the biochemical processes of ripening. But then, non-climacteric fruits seem to get along without it very well. Curious. Unlike non-climacteric fruits, mature climacteric types can be ripened when detached from the parent plant, a process that can be hastened by exposure to ethylene. For tissues responsive to it, ethylene is a potent inducer of ripening. It also triggers senescence of leaves and flowers in many (but not all) species.[257] Chemically or genetically blocking synthesis of, or responsiveness to, ethylene prevents fruit ripening. The little packets of chemicals that extend the life of cut flowers contain inhibitors of ethylene action. Unripe bananas will ripen if stored with ripe apples because apples produce lots of ethylene. For such a chemically simple little molecule, ethylene has a potent re-programming effect on many aspects of the development of many species.

According to the Food and Agriculture Organization of the United Nations, about one third of the edible fractions of food produced for human intake is wasted, amounting to 2.3 billion tons per annum. This equates to forfeiting one in every four calories intended for consumption. Food losses in industrialised countries are as high as in developing countries, but in the latter more than 40% of the food losses occur at the postharvest and processing stage, whereas in developed countries, more than 40% of losses occur at retail and consumer levels. Senescence and associated deteriorative processes contribute to both kinds of wastage. Senescence is often initiated and sustained by harvest and storage conditions. Postharvest technologies such as refrigeration, chemical treatments, storage in inert atmospheres and rapid transportation to the consumer can slow deterioration. Senescence and ripening from the point of retail onward are important factors in sell- and use-by dating, consumer perception and the implementation of commercial and legal quality standards. In the words of the World Resources Institute (WRI)[258] 'big inefficiencies suggest big savings opportunities.' In a 2013 report, WRI estimated that halving the current rate of food loss and waste would contribute 'roughly 22 percent of the 6,000 trillion kcal per year gap between food available today and that needed in 2050.' Tackling food loss and waste is identified by WRI to be an essential global strategy for achieving a sustainable food future. Understanding senescence and how to control it will make a sizeable contribution to achieving this goal.

Chapter 16 Loose ends

Pale creatures [259]

The yew endures four thousand years. The mayfly dances for a few summer hours.

And so it goes: every thing lives a lifetime.

Something lives under every stone – pale creatures hatching by the light of the moon.

And so it goes: every thing lives a lifetime.

Time, a hare and tortoise race. Lichen on a sundial's face.

Life's now. Time's never. And so it goes: the glass falls hour by hour, forever.

Daughters of Adam, sons of Eve, cast fleeting shadows on the walls of the cave.

And so it goes: every thing lives a lifetime.

⊢ 24 ⊣

Melting clockface, leafless tree. Apple where the head should be.

Still life. Still standing. And so it goes: to fly is easy, the hard part is landing.

Meanwhile, in awful solitude, out there two galaxies approach and collide.

And so it goes: every thing lives a lifetime

As we near the end of this journey through the world of senescence and ageing, what have we learned? That there are too many questions, and too few answers. Research on senescence addresses a whole range of almost intractable problems, some of which are beginning to yield to modern analytical approaches, others that may never be answered and will remain subjects for semantic or philosophical disputation. For example, there is the issue of definitions. An organism gets from a state of viability to death by any of a number of routes, many of which either do not involve physiological senescence at all, or in which senescence is a secondary or peripheral influence. By what criteria do we know that something is living, senescing, ageing or otherwise? Do we have definitions that allow us to say that an organism is really dead and not in diapause or some such suspended state? The comparative biology of ageing and senescence reveals a whole range of lifestyles. There are individuals such as lobsters and macrobian trees that, studies of demographic senescence appear to show, do not undergo ageing at all; creatures that age but usually die by accident (for example wild mice, which almost invariably get eaten before they have a chance to grow old); organisms such as monocarpic plants that exploit physiological senescence in order to age and die on their own terms; species such as humans that age and in which death is usually a consequence of ageing; and organs and organisms that can be rejuvenated (leaves, seeds, some invertebrates, for example). But what do we mean by an individual? Is a coral, or a sponge, or a slime-mould, or a tree a single organism or a population? There are special difficulties in trying to translate ageing behaviour from one level of biological organization to another.[260] The problem of scaling down from general to specific cases is one of these: What is the relationship (if any) between actuarial or demographic definitions of senescence and the behaviour of the individuals in the population? And how does any of this carry through to the level where mechanisms of senescence and related physiological processes are at work?

More questions, few answers. We see that all sorts of biological processes fail or decline with age, but which are causes that dictate the progress and nature of ageing, and which are merely symptoms? Here researchers encounter a problem in scaling up from the parts to the whole: Is there a relationship between senescence of component systems and that of the entire organism? This is sometimes called the 'One-Hoss Shay' effect, after the Oliver Wendell Holmes poem 'The Deacon's Masterpiece', in which the wonderful vehicle 'went to pieces all at once, and nothing first'.[261] Eyesight, blood pressure, joints, mental processes all degenerate with age in humans, but what has this to do with dying? In plants, what is the relationship between cell/tissue/organ death and survival of the whole organism? Is it meaningful to think in terms of a "master" reaction, a specific component the deterioration of which leads directly to whole-organism decline? Of particular significance for plants is the question of resource capture and allocation over time: is demographic senescence the symptom of a kind of starvation or neglect process and what does resource limitation mean for photosynthetic organisms, in which raw materials and energy are as abundant as light, air and water? What is the contribution of stressful environments to ageing? Does ageing literally result from being progressively more windswept, rugged and gnarly? How is the integration of a period of senescence into the full life cycle related to organism lifespan?

Finally there are questions about the mechanisms of ageing and senescence. Is ageing a failure of processes that normally defend against deterioration? What are the cost-benefit tradeoffs of repair, maintenance and durable construction? How is this related to the distinction between germline and soma in some organisms (though not plants)? Can ageing be avoided? Is ageing a failure to escape from influences that invoke the ageing response? Have organisms been able to channel the inevitability of ageing into processes that benefit their ecological and evolutionary fitness and what influence has this had on the programs for cell death and senescence? How can natural selection act to evolve genes with specific functions in ageing? Can 'ageing genes' be mutated, mapped and isolated? What are their environmental sensitivities? Can we ultimately disrupt ageing by tinkering with these genes? We have not yet properly framed many of these questions, let alone answered them. And even if the onward march of research provides more and more answers, senescence and ageing are likely to continue to provide grist to the philosopher's mill for as long as living beings remain mortal.

Chapter 17 Death

Doré's illustration of the crossing of the River Acheron from Canto III of Dante's
Inferno[262]

And so we arrive, finally, at the gates of Death itself. Death is the adjacent country to the Land of Biological Ageing and Senescence, but the life scientist (the clue is in the name) should approach the border with extreme caution. Ludwig Wittgenstein wrote 'Death is not an event in life'. Death is a condition or state and is the culmination of, and separate from, the process of dying – in the words of the Reverend Sydney Smith 'Death should be distinguished from dying, with which it is often confused'. Marcus Aurelius likened death to birth, calling it 'a secret of Nature'.[263] By definition, changes that occur in dead cells are post-mortem and non-biological. Biologists studying terminal events in development need to distinguish between the regulated activity of viable biological structures and the pathological outcomes of organic collapse. They do not always take such care.

...Then all together sorely wailing drew
To the curst strand, that every man must pass
Who fears not God. Charon, demoniac form,
With eyes of burning coal, collects them all,
Beckoning, and each that lingers, with his oar
Strikes. As fall off the light autumnal leaves,
One still another following, till the bough
Strews all its honours on the earth beneath...
    Dante. *Inferno*[264]

Senescing tissue is viable, dead tissue is not, and there is a transitional condition between the two states during which biochemistry modulates into abiotic chemistry. Writing about this some years ago, I thought it would be useful to have a term that described the twilight zone between the viable and the defunct, and found it in Robert Burton's *The Anatomy of Melancholy* (1621), where the author writes of 'an old acherontic dizzard, that hath one foot in the grave'. Acheron is the mythological river of Hades across which the newly dead are ferried by Charon in order to enter the Underworld.[265] The acherontic period in the lifespan of a tissue or organ is often rapid and always irreversible. During the preceding senescence phase, cell membranes and organelles remain intact, and there is no wilting because cells are viable and retain the capacity to regulate their state of hydration. In some cases, notably the photosynthetic tissues of leaves, senescence is under play-stop-rewind control until almost all salvageable material has been recycled and exported to the rest of the plant. The acherontic pathway leading from senescence to death is rather overlooked as an ageing-related process in its own right.[266] Which acherontic changes are symptoms and which are causes? For example, is vacuolar lysis in post-senescent leaf cells the agent of cell death or the consequence of lost viability? Is there such a condition as 'slightly dead'? Might ageing be a kind of slow-motion accumulation of acherontic events? How much of the active research area defined as Programmed Cell Death is really concerned with the acherontic state, and how many of the processes and mechanisms described are really post-mortem necrochemical changes? Questions, questions.

Inside the cell is a severe and stringent milieu. It has to be if viability is to be sustained. The fidelity and fitness of molecules, complexes and cell structures are continuously being tested by cytoprotective lytic systems that prowl the cell and pick off damaged, badly folded, mis-assembled, idle, inappropriate or superfluous components. The physiology of the cell is poised, like a tightrope-walker, inching along a potentially endless high-wire. For cells, the wire is time and in the end, poise will not be enough - the ropewalker will fall one way or the other and, in Shakespeare's words, be 'consumed with that which it was nourish'd by'. This, writ large, is one way we might picture ageing and death. Another, related, way is to think of ageing and death as inevitable consequences of the great biological principle underlying development, adaptation and survival of living organisms: namely, the closing down of options. The genome represents the impractical unedited totality of what the organism is capable of. The successful organisms are those that refrain from expressing inappropriate potential. The selectivity that orchestrates expression of genomic potential is exerted through cellular processes that repress and destroy. Maybe the doctrine of Original Sin isn't, after all, so utterly at odds with the the Enlightenment, Darwinism and the fading influence of Augustinian Christianity. Could it not be that ageing, and ultimately death, are the long-term revelations of these negative, but nonetheless essential, forces that animate the machinery of living matter?[267]

And with that, the biologist runs out of road. We can speak of Life, we can speak of Ageing, we can speak of Senescence. But biologists must leave Death to others. The rest is silence.[268]

Sources, resources and afterthoughts

[1] Digitised as a labour of love by Brandi Besalke. Online source: http://johnsonsdictionaryonline.com/?page_id=7070&i=1787 [accessed 23 February 2016]. The obscure word 'senectude', meaning old age, occurs in William Caxton's *Tulle of Old Age* (1481). *Senecio*, the systematic name of the groundsel genus, was so named by Pliny for the grey hairiness, reminiscent of that of an old man, that develops once fruiting begins.

[2] Samuel Johnson (1709-1784) was a large florid man with a large florid way with words. He had a cat called Hodge, whom he fed on oysters. The quotation is from James Boswell's *Life* (1791).

[3] Dashiell Hammett (1894-1961) was a thin, terse man who wrote tersely about very thin and very fat men. At one time he was a Pinkerton detective in San Francisco where, according to SJ Perelman (see note 163), he once met a woman called Dolores del Schultz. Lillian Hellman wrote that 'he knew the varieties of seaweed, and for a month he studied the cross-pollination of corn'.

[4] The field of senescence is cursed with semantic dispute. The plant scientist's vocabulary of ageing (US English: aging), life-history and senescence is (or should be) used in a different way from the gerontologist's, despite employing a number of common expressions. Off and on over the years I've tried to suggest how carefully defining terms might introduce some clarity to discussions of senescence, ageing and death (most recently in Thomas H. 2013. Senescence, ageing and death of the whole plant. New Phytologist 197: 696–711). The intention was not to be confrontational or prescriptive but simply to avoid misunderstandings and to establish as clearly as possible where it is profitable to seek common mechanisms and controls. This hasn't been universally welcomed (for example, van Doorn WG, Woltering EJ. 2004. Senescence and programmed cell death: substance or semantics? Journal of Experimental Botany 55: 2147–2153). In the present survey of the wider landscape of senescence, I've taken an uncharacteristically free and easy approach to the terminology surrounding the topic of senescence. I don't think it's done much harm, and it has certainly made writing a more relaxing business.

[5] The quotation is from Muir's poem 'The child dying'. The finest English translations of Franz Kafka's work were made by Muir and his wife Willa. Edwin Muir was born in 1887, but records in his diary for 1937-39 'I was born before the Industrial Revolution, and am now about two hundred years old.' He died in 1959.

[6] See Björn LO. 1976. *Light and Life* (Hodder and Stoughton). It's out of date, true, but the principles stand and are engagingly stated. The opening tall tale about the young Björn taking a pair of scissors to the flower pattern on his jumper is a delight. See also Weinberg S. 1983. *The First Three Minutes* (Fontana). The Epilogue, with its simultaneously bleak and uplifting conclusion, is a classic of popular science writing.

[7] J.C.A.B.M. Morton (1893-1979), English humourist, columnist and prankster. He was known to peer into postboxes, proclaiming 'Don't worry, little man, we'll soon have you out', and slipping away when a crowd had gathered. He and his friends used to leave empty beer bottles on Virginia Woolf's doorstep.

[8] We are all but recent leaves on the same old tree of life and if this life has adapted itself to new functions and conditions, it uses the same old basic principles over and over again. There is no real difference between the grass and the man who mows it (Albert Szent-Györgyi).

[9] I've tried as far as possible not to use a lot of heavy plant science shop-talk in these pages. Where jargon has been unavoidable, there's a definition of the technical term in the index. For a more detailed, but quite approachable, account of how plants are made and work, see Jones RL, Ougham HJ, Thomas H, Waaland S. 2013. *The Molecular Life of Plants* (Wiley).

[10] In 1975, the cultured and creative plant physiologist Aldo Carl Leopold published a thought-provoking paper titled 'Aging, Senescence, and Turnover in Plants' (BioScience 25: 659-662) in which he describes turnover as 'a transcendent theme in biology'. The paper is worth reading for the clarity of its arguments and its striking illustrations. One is a plot of turnover rates (measured as half-lives from 10^{-2} to 10^{10} days) at each level of biological organisation, from molecules at one extreme, through individual organisms to whole floras at the other. The paper was written at the dawn of the present era of molecular analysis, and the author, while acknowledging the significance of the genetic programming of senescence, was unable to see how the state of knowledge at the time allowed the link between turnover and senescence to be defined. He makes the intriguing proposition, however, that ageing may be a kind of a fail-safe mechanism essential for assuring turnover in biological systems takes place.

[11] In a book about senescence, it's appropriate to quote T.S. Eliot (1888-1965), author of poems entitled 'Gerontion' and 'Whispers of Immortality'. In his biography of the poet's early years, *Young Eliot: From St Louis to The Waste Land* (Jonathan Cape, 2015), Robert Crawford writes that Eliot, who had to wear a truss all his life, was 'never young'.

[12] Harrison RP. 2014. *Juvenescence: A Cultural History of Our Age* (University of Chicago Press).

[13] Harrison (2014).

[14] W.B. Yeats 'The Second Coming' (1919).

[15] Schrödinger E. 1944. *What is life? The Physical Aspect of the Living Cell* (University of Michigan Press).

[16] A beautiful quotation from the great diarist, essayist, critic and eroticist Anaïs Nin (Angela Anaïs Juana Antolina Rosa Edelmira Nin y Culmell; 1903-1977).

[17] Sigmoidal curves generally belong to the logistic function family. My late friend and colleague David Causton was an early adopter of computerised logistic curve fitting for plant growth analysis – see Causton DR, Venus J. 1981. *The Biometry of Plant Growth* (Edward Arnold). I have done a bit of this in my time (and very satisfying it was, too); for example Thomas H, Potter JF. 1985. Fitting logistic type curves to extension growth data for leaves of grass species by means of the Maximum Likelihood Program: analysis of leaf extension in *Lolium temulentum* at optimal and chilling temperatures. Environmental and Experimental Botany 25: 157–163.

[18] See Zhang J, Nieminen K, Serra JAA, Helariutta Y. 2014. The formation of wood and its control. Current Opinion in Plant Biology 17: 56-63.

[19] The following papers on stomata provide evidence that the cells of different plant tissues undergo senescence according to different timetables: Zeiger E, Schwartz A. 1982. Longevity of guard cell chloroplasts in falling leaves: implication for stomatal function and cellular aging. Science 218: 680-682; and Keech O, Pesquet E, Gutierrez L, Ahad A, Bellini C, Smith SM, Gardeström P. 2010. Leaf senescence is accompanied by an early disruption of the microtubule network in Arabidopsis. Plant Physiology 154: 1710-20.

[20] Finch CE. 1994. *Longevity, Senescence, and the Genome* (University of Chicago Press).

[21] See Thomas H, Huang L, Young M, Ougham H. 2009. Evolution of plant senescence. BMC Evolutionary Biology 9: 163 (doi:10.1186/1471-2148-9-163).

[22] Richard Milhous Nixon (1913-1994), disgraced 37th President of the United States of America. To my surprise I found him listed in a book of Americans of Welsh ancestry. It seems he is descended on his mother's side from landed gentry in the Wrexham area.

[23] Karl Marx (1818-1883), luxuriantly bearded economist, philosopher and revolutionary socialist. Strangely, the arch-critic of capitalism did not use the word (*Kapitalismus*) at all in volume I (1867) of *Das Kapital* (volumes II and II were published posthumously by Friedrich Engels in 1885 and 1894 respectively).

[24] For a discussion of canopy turnover, see Hikosaka K. 2005. Leaf canopy as a dynamic system: ecophysiology and optimality in leaf turnover. Annals of Botany 95: 521-533.

[25] The concept of modular organisation in biology is generally attributed to William Bateson (1861-1926), an early champion of the ideas of Mendel and originator of the word 'genetics'. The body plan of multicellular organisms consists of repeated segments (metamers or somites). Earthworms clearly display this mode of construction, and even vertebrates (including humans) are, deep down, metamerically segmented in

developmental and anatomical organisation. The structural module in plants is called the phytomer (Rutishauser R, Sattler R. 1985. Complementarity and heuristic value of contrasting models in structural botany. Botanische Jahrbücher für Systematik, Pflanzengeschichte und Planzengeographie 107: 415-455). The phytomer is the basic unit for mathematical modelling of plant structure and development (Vos J, Evers JB, Buck-Sorlin GH, Andrieu B, Chelle M, de Visser PHB. 2010. Functional–structural plant modelling: a new versatile tool in crop science. Journal of Experimental Botany 61: 2101-2115).

[26] An example of scaling up from the individual to the biome is GlobAllomeTree, an initiative of the Food and Agriculture Organization of the United Nations (FAO) to apply the principles of allometric analysis to assess forest volume, biomass and carbon stocks: see Picard N, Saint-André L, Henry M. 2012. *Manual for building tree volume and biomass allometric equations: from field measurement to prediction.* (Food and Agricultural Organization of the United Nations, Rome, and Centre de Coopération Internationale en Recherche Agronomique pour le Développement, Montpellier). Allometry is the science of the scaling relationship between the size of a body part and the size of the body as a whole during growth and development. It models the functional and evolutionary mechanisms by which such relationships determine morphological, physiological and ecological traits.

[27] Klarsfeld A, Revah F. 2004. *The Biology of Death: Origins of Mortality* (Cornell University Press).

[28] August Friedrich Leopold Weismann (1834-1914) was a German biologist and early supporter of Darwin's theory of natural selection. A collection of his writings, published in English as *Essays upon Heredity and Kindred Biological Problems* (1889), included discussions of senescence (arguing that it is the outcome of natural selection, allowing adapted organisms to reach reproductive maturity), the inheritance of acquired characteristics, and (his most celebrated achievement) the germ-plasm theory. To test Lamarckism and the germline-soma principle, he cut off the tails of 901 mice and their offspring for five generations and showed that the amputees' descendants still grew tails like normal mice. The farmer's wife's experiment with three visually impaired rodents probably would have benefitted from more replication.

[29] With so many words to choose from, it seems a pity to pick on one that already had a perfectly clear botanical meaning in the *ninth century*. Even plant scientists have started using 'stem cell' to mean meristematic or undifferentiated cell. This kind of surrender to the imperialism of biomedical terminology is depressing for those of us who respect and try to preserve botanical tradition.

[30] See Kirkwood TB. 1977. Evolution of ageing. Nature 170: 201-204. Kirkwood continues to argue against programming as the basis of ageing (for example Kirkwood T, Melov S. 2011. On the programmed/non-programmed nature of ageing within the life history. Current Biology 21: R701-707).

[31] Aesthetically and physiologically meaningful virtual plants have been generated using fractal mathematics. For a beautiful account of the origins and early achievements of the recursive approach to modelling plant forms, see Prusinkiewicz P, Lindenmayer A. 1990. *The Algorithmic Beauty of Plants* (Springer-Verlag).

[32] I've discussed this: Thomas (2013) – see note 4.

[33] In *Structures: or why things don't fall down* (Penguin, 1991).

[34] Schrödinger (1944) – see note 15.

[35] Kurt Vonnegut Jr. (1922-2007), counterculture novelist. The quotation is well known, but the source isn't very clear. One suggestion is that it comes from 'A Letter to the Next Generation' in an Open Forum series of advertisements sponsored by Volkswagen and printed in *Time* magazine, February 1988. Or maybe not. So it goes.

[36] Don Van Vliet (1941-2010), musician and artist. In the words of John Peel 'A psychedelic shaman who frequently bullied his musicians and sometimes alarmed his fans'.

[37] Brian Howard Clough (1935-2004), footballer, club manager and controversialist.

[38] From the Greek *Thanatos*, the personification of death and twin brother of *Hypnos* (sleep). Thanatology is the study of death. On 1 March 1972 the *Daily Colonist* (Victoria, B.C.) reported that the 450 members of the Corporation of Funeral Directors and Embalmers of Quebec had renamed their organisation the Corporation of Thanatologists. The term 'athanogene' has been coined to describe a gene that confers resistance to apoptosis (cell death).

[39] Remote sensing shows that, each year, the wave of autumnal chlorophyll loss advances north to south at a rate of about one degree of latitude every 4 days (more than 1 kilometre per hour). As you would expect for a phenomenon visible from space, global chlorophyll turnover consists of some big numbers. We can do interesting calculations using data given in the ground-breaking review by Hendry GAF, Houghton JD, Brown SB. 1987. The degradation of chlorophyll - a biological enigma. New Phytologist 107: 255-302. About 288 billion kilograms of terrestrial chlorophyll is broken down each year. The vegetation of the British Isles contains around 473 million kilograms of chlorophyll. Assuming a global average rate of turnover, total degradation of all the chlorophyll in Britain would take less than an hour and a half. At the time George Hendry and colleagues were drawing attention to the global significance of chlorophyll breakdown, we had a pretty detailed idea of the biochemical pathway for making chlorophyll but knew nothing about the other side of the turnover equation, how chlorophyll is unmade. Thirty years later, from a standing start, we are now able to explain seasonal changes in satellite images and global-level pigment dynamics right down to the level of cells, organelles, enzymes and genes.

40 John Ruskin (1819-1900), art critic, social commentator and champion of gothic architecture – a dramatic example of which, the Old College Building (designed by John Nash, built in the 1860s) on the sea at Aberystwyth, I can see from my window. The story that Ruskin's marriage to Effie Gray wasn't consummated because of his shock and revulsion on the wedding night at discovering that she, unlike the classical statues that were his only experience of the female form, had pubic hair is now pretty well disproved. Whatever it was that put Ruskin off this celebrated beauty remains a mystery.

41 John Keats (1795-1821), Percy Bysshe Shelley (1792-1822), leading romantic poets. To judge by their short and sweet lives, creating romantic poetry must have been a high-risk occupation. See the following books by my friend and colleague Richard Marggraf Turley: *Keats's Boyish Imagination* (Routledge, 2003) and *Bright Stars: John Keats, Barry Cornwall and Romantic Literary Culture* (Liverpool University Press, 2009).

42 British Journal of Cancer 26: 239–57.

43 McCormack ML, Eissenstat DM, Prasad AM, Smithwick EAH. 2013. Regional scale patterns of fine root lifespan and turnover under current and future climate. Global Change Biology 19, 1697–1708. Tree root lifespans range from 2 to 3 weeks (apple) to 2 to 3 years (sycamore). Estimates of the proportion of net primary productivity allocated to turnover of fine roots vary from 10% to more than 40%. Root senescence makes a global-scale contribution to the burial of carbon in the soil – one of the most effective counter-measures against the inexorable rise in atmospheric greenhouse gas concentration.

44 The term 'mycorrhiza' (from the Greek 'fungus' and 'root') was first used by the German botanist Albert Bernhard Frank (1839-1900). There's an English translation of Frank's classic 1885 paper here: http://tinyurl.com/ox5sofe [accessed 23 February 2016]. It has been estimated that 80% of plant species form relationships with mycorrhizal fungi. It's a pity that Arabidopsis, the model organism universally used by researchers for genetic and physiological analysis of plant development, is one of the 20% of species that are non-mycorrhizal. Mycorrhizal associations are generally beneficial to both fungi and plant hosts. The two most abundant types are the arbuscular mycorrhizas, which penetrate cells of the host, and the ectomycorrhizas, in which the association is extracellular. The standard reference is the 3rd (2010) edition of *Mycorrhizal Symbiosis*, by SE Smith and DJ Read (Academic Press).

45 Yosa Buson (1716-1784), Japanese poet and painter.

46 This song, a favourite of jazz singers (recordings by Blossom Dearie and Nancy Wilson are particularly highly recommended), is from the 1954 Broadway musical *House of Flowers*, based on a short story by Truman Capote (1924-1984). The original production is said to be the first time outside Trinidad and Tobago that the steel pan was heard in the theatre.

Harold Arlen (1905-1986) was a great composer of many classic songs, and wrote the music for *The Wizard of Oz*.

[47] The Hungarian-French composer Joseph Kosma (1905-1969) composed 'Les feuilles mortes' as a pas de deux for the 1945 ballet *Le Rendez-vous*. The original lyrics were by the poet and screenwriter Jacques Prévert (1900-1977). The English words were written by Johnny Mercer (1909-1976) on a short train journey to New York in 1950. 'Autumn leaves' is regarded as the most important non-American standard popular song and has been recorded more than fourteen hundred times. It also serves as the Pons Asinorum for the student of jazz harmony and improvisation.

[48] Original publications identifying the cluster of genes for senescence, stature and photoperiod sensitivity in wheat include: Pestsova E, Röder M. 2002. Microsatellite analysis of wheat chromosome 2D allows the reconstruction of chromosomal inheritance in pedigrees of breeding programmes. Theoretical and Applied Genetics 106: 84-91; Verma V, Foulkes MJ, Worland AJ, Sylvester-Bradley R, Caligari PDS, Snape JW. 2004. Mapping quantitative trait loci for flag leaf senescence as a yield determinant in winter wheat under optimal and drought-stressed environments. Euphytica 135: 255–263.

[49] For the challenges that the Green Revolution sought (and still seeks) to meet, see Evans LT. 1998. *Feeding the Ten Billion: Plant and Population Growth* (Cambridge University Press). The 'father of the Green Revolution', Norman Borlaug (1914-2009), was awarded the Nobel Peace Prize in 1970. Some green activists try to demonise him. Ignore them – his great humanitarian achievements will stand as an enduring rebuke to the vinegary drizzle of their ill-intentioned carping.

[50] Pieter Cornelis Mondriaan (1872-1944), Dutch artist. Until he moved to Paris in 1911 he painted many trees and landscapes in a naturalistic or impressionistic style, but seems to have been averse to the colour green from an early stage. Green is completely absent from the Neo-plasticism (geometric non-representational works) of his mature style. Other members of the *De Stijl* movement, with which Mondrian was associated, also avoided green, as did Wassily Kandinsky ('Green is like a fat, very healthy cow lying still and unmoving, only capable of chewing the cud, regarding the world with stupid dull eyes' – *Concerning the Spiritual in Art*, 1910). Elsewhere, Jayne Elisabeth Archer, Richard Marggraf Turley and I have discussed how these attitudes of rejection arose in the twentieth century, and their social and cultural implications (*Food and the Literary Imagination*, Palgrave, 2014).

[51] James Joyce has a quotation for every occasion. This is from *Ulysses*, Episode 12 ('Cyclops'), in which Leopold Bloom encounters The Citizen and his dog Garryowen.

[52] For an accessible account of what goes on in a laboratory that carries out research on Arabidopsis, see Harberd N. 2012. *Seed to Seed: The Secret Life of Plants* (Bloomsbury).

[53] See: Keskitalo J, Bergquist G, Gardeström P, Jansson S. 2005. A cellular timetable of autumnal senescence in aspen. Plant Physiology 139: 1635-1648; and Jansson S, Thomas H. 2008. Senescence: developmental program or timetable? New Phytologist 179: 575–579.

[54] Chapter 8 of *The Molecular Life of Plants* (see note 9) gives an overview of how plants perceive and respond to different wavelengths of light.

[55] Phytochrome exists in two photoreversible forms. P_R absorbs red light (wavelength around 660 nanometres) and is converted to P_{FR}, which reverts to P_R when exposed to far-red light (about 730 nanometres). If a developmental process is red/far-red reversible (an example is the light-sensitive germination of lettuce seed), it's probably mediated by phytochrome. The phytochrome protein is attached to a photoreactive molecule that toggles between two structural forms in response to red or far-red light. Photoconversion from P_R to P_{FR} opens up the phytochrome protein, making buried surfaces accessible to interacting factors that in turn initiate physiological responses. Phytochrome enables plants to perceive the state of the light environment, including photoperiod, by sensing changes in the ratio of red to far-red wavelengths. For the full phytochrome story, see Sage LC. 2012. *Pigment of the Imagination: A History of Phytochrome Research* (Elsevier).

[56] English poet, born 1852, died 1921. 'The night has a thousand eyes', published in 1890, is his best-known poem and a phrase that has taken on a life of its own, not least as the title (and title song) of a 1948 film-noir starring Edward G. Robinson, and an inferior pop song performed in 1965 by the teen idol Bobby Vee.

[57] See Franklin KA, Whitelam GC. 2004. Light signals, phytochromes and cross-talk with other environmental cues. Journal of Experimental Botany 55: 271-276. As well as phytochrome, this review discusses other photoreceptor types, notably cryptochromes and phototropins.

[58] For more on this subject, see Kebrom TH, Brutnell TP. 2007. The molecular analysis of the shade avoidance syndrome in the grasses has begun. Journal of Experimental Botany 58: 3079-3089. One of the most creative of plant physiologists, the late Harry Smith (1935-2015), suggested that agricultural productivity could be improved without increased demand for agrochemicals by a strategy of eliminating proximity perception in crop plants through genetic manipulation of phytochrome genes (Smith H. 1992. The ecological functions of the phytochrome family. Clues to a transgenic programme of crop improvement. Photochemistry and Photobiology 56: 815–822).

[59] Hutchinson TC. 1967. Comparative studies of the ability of species to withstand prolonged periods of darkness. Journal of Ecology 55: 291–299.

[60] Read J, Francis J. 1992. Responses of some southern hemisphere tree species to a prolonged dark period and their implications for high-latitude Cretaceous and Tertiary floras. Paleogeography Paleoclimatology and Paleoecology 99: 271–290.

[61] Writer, born 1835, died 1902. The quotation is from the *Notebooks* (1917). Butler's best-known work is *Erewhon* which, among other things, predicted the rise of Artificial Intelligence. He made translations of Homer and became convinced that *The Odyssey* had been written by a young Sicilian woman.

[62] For example: Vijayalakshmi K, Fritz AK, Paulsen GM, Bai G, Pandravada S, Gill BS. 2010. Modeling and mapping QTL for senescence-related traits in winter wheat under high temperature. Molecular Breeding 26: 163-175; Lobell DB, Sibley A, Ortiz-Monasterio JI. 2012. Extreme heat effects on wheat senescence in India. Nature Climate Change 2: 186–189.

[63] Studies of many plant species have discovered genetic loci related to temperature and drought responses coincident with loci in the genome that govern leaf senescence behaviour (see Ougham H, Armstead I, Howarth C, Galyuon I, Donnison I, Thomas H. 2007. The genetic control of senescence revealed by mapping quantitative trait loci. Annual Plant Reviews 26: 171-201). Sorghum represents perhaps the most successful example of improving crop stress tolerance by targeting senescence. For the genetic and physiological background to this strategy, see Thomas H, Ougham H. 2014. The stay-green trait. Journal of Experimental Botany 65: 3889-3900. My friend and colleague Andy Borrell is currently one of the leaders of an international effort to turn the strategy into practical outcomes for agriculture in the semi-arid regions of the developing world (Borrell AK, van Oosterom EJ, Mullet JE, George-Jaeggli B, Jordan DR, Klein PE, Hammer GL. 2014. Stay-green alleles individually enhance grain yield in sorghum under drought by modifying canopy development and water uptake patterns. New Phytologist 203: 817–830).

[64] The distinctive holes in the leaves of the Swiss cheese plant arise by the selective programmed death of groups of cells early in development (Gunawardena AHLAN, Sault K, Donnelly P, Greenwood JS, Dengler NG. 2005. Programmed cell death and leaf morphogenesis in *Monstera obliqua* (Araceae). Planta 221: 607-618). The aquatic species *Aponogeton madagascariensis* is called the lace plant for the beautiful tracery of perforations in its leaves, the development of which is also the consequence of localised programmed death of cells in the newly-formed organ (Lord CE, Gunawardena AH. 2011. Environmentally induced programmed cell death in leaf protoplasts of *Aponogeton madagascariensis*. Planta, 233: 407-421).

[65] One of the genes controlling elongation growth in deep-water rice is the appropriately named *SNORKEL* (*SK*). Another is *SUBMERGENCE* (*SUB*). Leaf senescence under prolonged darkness is delayed in rice carrying the submergence tolerance gene (Fukao T, Yeung E, Bailey-Serres J. 2012. The submergence tolerance gene *SUB1A* delays leaf senescence under prolonged darkness through hormonal regulation in rice. Plant Physiology 160: 1795–1807). Transferring *SK* or *SUB* genes into productive varieties that are normally flooding-intolerant has the prospect of improving rice yield and quality.

[66] The agronomic importance of aerenchyma has been surveyed by Yamauchi T, Shimamura S, Nakazono M, Mochizuki T. 2013. Aerenchyma formation in crop species: a review. Field Crops Research 152: 8-16. Aerenchyma as an example of programmed senescence of specific cells is discussed by Gunawardena A, Pearce DM, Jackson MB, Hawes CR, Evans DE. 2001. Characterisation of programmed cell death during aerenchyma formation induced by ethylene or hypoxia in roots of maize (*Zea mays* L.). Planta 212: 205–214.

[67] On this subject, as on so many others, it is worth reading D'Arcy Wentworth Thompson (1860-1948). Scale and the question of surface area and volume are covered in Chapter II of his magisterial *On Growth and Form* (1917; Dover reprint 1992). Not the least remarkable feature of this wholly extraordinary work is that in all its 1116 pages there is not a single mention of genes and genetics, even though it is foundational for the modern study of morphogenesis. Thompson was a representative of a vanished breed – the erudite, classically educated, polymathic natural scientist. He also fully lived up to the expectations of professorial eccentricity, walking the streets of St Andrews, where he held the University Chair of Natural History, with a parrot on his shoulder.

[68] Simcha Lev-Yadun has discussed some of the rare examples of cell movement during plant development (Lev-Yadun S. 2001. Intrusive growth – the plant analog of dendrite and axon growth in animals. New Phytologist 150: 508–512).

[69] This is the earliest written occurrence of the word 'zest', according to the *Oxford English Dictionary* (online edition 2013, 2015 update).

[70] Peterson C, Seligman MEP. 2004. *Character Strengths and Virtues: a Handbook and Classification* (Oxford University Press).

[71] See Liu P, Liang S. Yao N, Wu H. 2012. Programmed cell death of secretory cavity cells in fruits of *Citrus grandis* cv. Tomentosa is associated with activation of caspase 3-like protease. Trees 26: 1821-1835.

[72] Lysigeny and schizogeny are discussed by Turner GW. 1999. A brief history of the lysigenous gland hypothesis. Botanical Review 65: 76-88, and Pickard WF. 2008. Laticifers and secretory ducts: two other tube systems in plants. New Phytologist 177: 877-888. 'Lysigeny' should not be confused with 'lysogeny', one of the cycles of viral replication in a bacterial host cell (Bertani G. 2004. Lysogeny at mid-twentieth century: P1, P2, and other experimental systems. Journal of Bacteriology 186: 595-600). 'Schizogeny' was the title of Episode 9 in Season 5 of the ludicrous but somehow compulsively viewable TV series 'The X-Files'.

[73] In *The Killing Room* (dir. Jonathan Liebesman, 2009), a straight-to-DVD horror-thriller that has slowly gained cult status.

[74] In 1963, the great Sydney Brenner (born 1927; Nobel laureate 2002) suggested that *C. elegans* would make a suitable model animal with simple body plan for molecular studies of development. Since then, it has become one of the most important subjects for ageing research (Tissenbaum HA. 2015. Using *C. elegans* for aging research. Invertebrate Reproduction and Development 59, supplement 1: 59-63). Normally *C. elegans* has a lifespan in the laboratory of about 17 days. Some mutants live twice as long as this. A large number of mutants and genetically-manipulated variants have been described in which the fates of individual cells have been modified, with implications for the longevity of the organism and its components. In some cases, the corresponding *C. elegans* genes have been related to genes with similar functions in humans and other animals. Apoptosis and autophagy are types of cell death in *C. elegans* and humans that share common molecular and cell biological mechanisms. In the pantheon of model organisms for molecular genetics research, *C. elegans* is 'the Worm' (*Drosophila* is 'the Fly'; Arabidopsis is 'the Weed').

[75] Earl 'Fatha' Hines (1903-1983), influential transitional figure between the classic jazz piano approach of the 1920s and 30s and the modern style of the 1940s and 50s. For a while around 1943 he led a big band that included Charlie Parker and Dizzy Gillespie, an experience that severely tried his patience. By the early 60s he was an all-but-forgotten figure from jazz history. I spent some time in the mid-70s at the University of California in Berkeley, and one of the people in the lab, knowing of my interest in jazz, mentioned that Hines was a neighbour of his, living quietly in Oakland. This was how I got to speak with this living legend, who was unnecessarily grateful to meet someone that remembered his achievements. So far as I could tell, his hands were perfectly normal.

[76] Syndactyly has an incidence of around 1 in 2000 live births and is twice as common in males, as well as in the Caucasian population (Jordan D, Hindocha S, Dhital M, Saleh M, Khan W. 2012. The epidemiology, genetics and future management of syndactyly. The Open Orthopaedics Journal 6: 14-27). Suppression of apoptosis results in retention of webbing between digits: see Abud HE. 2004. Shaping developing tissues by apoptosis. Cell Death and Differentiation 11: 797–799. It has been frequently stated (but never verified) that Josef Stalin had webbed toes.

[77] An astonishing organism! Photograph of *Welwitschia mirabilis* by Thomas Schoch. Licensed under CC BY-SA 3.0 via Wikimedia Commons - http://tinyurl.com/myqlqpl [accessed 8 May 2015].]. *Welwitschia* was described by, and named for, the Austrian botanist Friedrich Martin Josef Welwitsch (1806-1872). The core of the plant is edible (the 'onion of the desert') and is part of the diet of the indigenous people of the Namib.

[78] George Gordon Noel Byron, 6th Baron Byron (1788-1824). He turned up in Switzerland following flight into exile from London to escape bailiffs, scandals galore and a failed marriage. So full-bloodedly did he play the role of Bad Boy Romantic Poet that it was not until 145 years after his death that Westminster Abbey relented and allowed him a place in Poets' Corner.

[79] Mary Wollstonecraft Shelley (1797-1851). Her earliest published work was the poem 'Mounseer Nongtongpaw', written when she was 10 years old. The first of the very many (at least 60) motion picture versions of *Frankenstein* was made in 1910 by Thomas Edison. It is 12 minutes long.

[80] Born 1803, died 1873. Inevitably, he was described as 'hopelessly useless' by his schoolmaster, before going on to become one of the great innovative chemists of his era, and to invent that precious gift to half of the human race, Marmite.

[81] The standard work on plant nutrients is *Marschner's Mineral Nutrition of Higher Plants* (2012 edition, Academic Press). It covers everything from the classification, deficiency symptoms and uptake of essential minerals to their roles in physiology (including senescence), yield, soil-plant relations and global geochemical cycles. For a thorough analysis of nutrient remobilisation, deficiency and senescence in different species, see Maillard A, Diquélou S, Billard V, Lainé P, Garnica M, Prudent M, Garcia-Mina J, Yvin J, Ourry A. 2015. Leaf mineral nutrient remobilization during leaf senescence and modulation by nutrient deficiency. Frontiers in Plant Science 6: 317.

[82] The nitrogen experiment was carried out in the laboratory of the illustrious plant physiologist Kenneth V Thimann (1904-1997). Oat plants grown from seed without an external supply of nitrogen flowered after forming only 4 leaves and each set a single grain (Mei H, Thimann KV. 1984. The relation between nitrogen deficiency and leaf senescence. Physiologia Plantarum 62: 157–1610). Helen Ougham and Mair Lloyd Evans obtained a similar result with *Lolium temulentum* grown in the absence of phosphorus. These observations make you think. A large proportion of the phosphorus and nitrogen used and re-used by the plants in these experiments will have been in the form of DNA and RNA. For plants, nucleic acids are more than repositories of genetic information; they are caches of essential nutrients. DNA, being chemically stable and safely packaged within the cell in chromosomes and the nucleus, makes an excellent phosphorus and nitrogen stockpile. It has been argued that this is an energetically inefficient way to lay down reserves; but what does 'energy efficiency' mean for an organism that's awash with energy every time the sun comes up? It's not too outlandish to imagine that nutrient hoarding accounts in part for why plant genomes vary enormously in size, between even quite closely related species, as the result of replicative processes that confer developmental and ecological flexibility. See Grime JP, Mowforth MA. 1982. Variation in genome size—an ecological interpretation. Nature 299: 151-153, and Schole, DR, Paige KN. 2015. Plasticity in ploidy: a generalized response to stress. Trends in Plant Science 20: 165–175.

[83] Named for the German chemists and Nobel laureates Fritz Haber (1868-1934) and Carl Bosch (1874-1940). It is estimated that almost 80% of the nitrogen found in human tissues has come through the Haber-Bosch process.

[84] Plant species (both legumes and non-legumes) and their nitrogen-fixing symbionts are listed in Table 12.2 of *The Molecular Life of Plants* (see note 9).

[85] See note 44.

[86] For some figures on the carbon costs of making and maintaining nodules and fixing nitrogen, see Voisin AS, Salon C, Jeudy C, Warembourg FR. 2003. Symbiotic N_2 fixation activity in relation to C economy of *Pisum sativum* L. as a function of plant phenology. Journal of Experimental Botany 54: 2733-2744.

[87] Charpentier M, Oldroyd G. 2010. How close are we to nitrogen-fixing cereals? Current Opinion in Plant Biology 13: 556–564. One of us is wrong...

[88] Note 44

[89] This leads one to expect that soils of the inner regions of landmasses remote from the ocean will be relatively phosphorus-depleted compared with coastal regions. Looking at maps of global phosphorus abundance, I can imagine this might be broadly true, at least for Africa and South America – see Xiaojuan Y, Post WM, Thornton PE, Jain A. 2013. The distribution of soil phosphorus for global biogeochemical modeling. Biogeosciences 10: 2525-2537.

[90] See Cordell D, Drangert J, White S. 2009. The story of phosphorus: global food security and food for thought. Global Environmental Change 19: 292-305.

[91] A consortium of ecologists has brought together a huge database of leaf traits for 2548 species from across 175 sites (Wright IJ et al. 2004. The worldwide leaf economics spectrum. Nature 428: 821-827). The range of leaf lifespans varied from 0.9 up to 288 months. Leaf nitrogen content (per unit mass) ranged from 0.2% to 6.4%, and phosphorus from 0.008% to 0.6%. Leaf lifespan was highly negatively correlated with nitrogen and phosphorus contents and photosynthetic capacity. The authors interpret these relations to mean that long leaf lifespans require the robustness and low palatability that deter herbivores. High photosynthetic capacity, high nitrogen content and short lifespans tend to go together because they are associated with increased vulnerability to herbivory and physical hazards. High photosynthesis also drives fast growth, leading to shading, senescence and nutrient recovery from older leaves. By linking global-scale databases of traits and modelling approaches, ecologists are beginning to move between scales, relating the life-histories of plant forms to those of their structural components. See, for example Adler PB, Salguero-Gómez R, Compagnoni A, Hsu JS, Ray-Mukherjee J, Mbeau-Ache C, Franco M. 2014. Functional traits explain variation in plant life history strategies. Proceedings of the National Academy of Sciences 111: 740-745.

[92] The lifespan of heather leaves is strongly reduced by applying NPK fertiliser (Aerts R. 1989. The effect of increased nutrient availability on leaf turnover and aboveground productivity of two evergreen ericaceous shrubs. Oecologia 78: 115-120).

[93] Ewers FW, Schmid R. 1981. Longevity of needle fascicles of *Pinus longaeva* (bristlecone pine) and other North American pines. Oecologia 51: 107-115.

[94] This estimate of leaf age in *Welwitschia* is a bit misleading. The plant has two strap-like leaves which grow continuously from the base, and a millennium-old individual will have retained the same two leaves throughout its lifespan. Nevertheless, cell viability declines towards the leaf tip, which becomes frayed and moribund. It is estimated that a given cohort of leaf cells might survive for 1 to 10 years from formation at the base of the leaf to death at the tip. This is an example of the Ship of Theseus paradox – Plutarch, in his *Life of Theseus*, described how the vessel had been restored by replacing over time every one of its wooden parts, and posed the question, is it then the same ship? Is the leaf of a *Welwitschia* individual the same leaf after 10, 100, 1000 years? For a comprehensive account of the growth of *Welwitschia*, and a useful bibliography, see Henschel JR, Seely MK. 2000. Long-term growth patterns of *Welwitschia mirabilis*, a long-lived plant of the Namib Desert (including a bibliography). Plant Ecology 150: 7-26.

[95] The quotation is from *A Wizard of Earthsea* (1968).

[96] A particular hazard when plants are exposed to strong light is photoinhibition, loss of photosynthetic capacity due to damage to the reaction centre that splits water and produces oxygen. See Murata N, Takahashi S, Nishiyama Y, Allakhverdiev SI. 2007. Photoinhibition of photosystem II under environmental stress. Biochimica et Biophysica Acta – Bioenergetics 1767: 414–421.

[97] The earliest use of the term 'pseudo-senescence' I have found is in connection with area-volume relations in invertebrate growth (Bidder GP. 1932. Senescence. British Medical Journal 2: 583-585).

[98] For a discussion of the differences between true and pseudo-senescence, see Ougham H, Hörtensteiner S, Armstead I, Donnison I, King I, Thomas H, Mur L. 2008. The control of chlorophyll catabolism and the status of yellowing as a biomarker of leaf senescence. Plant Biology 10: 4–14.

[99] To Shakespeare they were 'chimney sweepers'. For John Clare they are 'nothing else but down, while the rude winds blow off each shadowy crown'. But for most people, and children everywhere, they are dandelion clocks. Illustration, by Debbie Maizels, from Jones et al. 2013. *The Molecular Life of Plants* (see note 9), with permission.

[100] Vladimir Vladimirovich Nabokov (1899-1977), writer, entomologist, uncompromising opponent of Freud (the 'Viennese witchdoctor') and Darwin. The quotation is from his autobiography *Speak, Memory* (1951).

[101] From the satire *The Staple of News* (1626). Jonson (1572-1637) was rival, critic and admirer of 'My Beloved the Author, Mr. William Shakespeare'.

[102] See note 53.

[103] Different measures of time related to plant physiology are discussed by Trudgill DL, Honek A, Li D, van Straalen NM. 2005. Thermal time – concepts and utility. Annals of Applied Biology 146: 1–14 and Meicenheimer RD. 2014. The plastochron index: Still useful after nearly six decades. American Journal of Botany 101: 1821-1835.

[104] Hayflick (born 1928) is a leader in the field of the cellular basis of ageing. He showed that cultured animal cells are not immortal. He is strongly opposed to 'anti-ageing medicine'. His 1998 publication, titled 'Aging is not a disease' (Aging [Milan] 10: 146), says it all.

[105] For an account of the relationship between the Hayflick limit, telomere attrition and ageing, see Bernadotte A, Mikhelson VM, Spivak IM. 2016. Markers of cellular senescence. Telomere shortening as a marker of cellular senescence. Aging (Albany, NY) 8: 3-11.

[106] The following publication describes the formation of a Europe-wide network to monitor phenological change: van Vliet, AJ et al. 2003. The European phenology network. International Journal of Biometeorology 47: 202-212. A useful website, with links to datasets (including senescence observations) freely available for further analysis, is: www.naturescalendar.org.uk/research/phenology.htm [accessed 23 February 2016].

[107] Fossil evidence for the differentiation of vascular systems and the shedding of parts in the first land plants suggests that lysigeny and schizogeny are ancient traits - see Edwards D, Li C-S, Raven JA. 2006. Tracheids in an early vascular plant: a tale of two branches. Botanical Journal of the Linnaean Society 150: 115-130; and Edwards D. 1993. Cells and tissues in the vegetative sporophytes of early land plants. New Phytologist 125: 225–247.

[108] Helen Ougham, Lin Huang, Mike Young and I have surveyed the evolutionary origins of senescence – see note 21.

[109] Animator, film maker, crooner and activist against anti-science attitudes ('The resistance to science is idiotic. Those people shouldn't be allowed to have antibiotics. Give us back your TVs and the dentures').

[110] See Thomas et al. (2009) – note 21.

[111] See Helferich G. 2004. *Humboldt's Cosmos: Alexander Von Humboldt and the Latin American Journey that Changed the Way We See the World* (Gotham Books).

[112] See, for example, Feild TS, Arens NC. 2005. Form, function and environments of the early angiosperms: merging extant phylogeny and ecophysiology with fossils. New

Phytologist 166: 383–408; Rausher MD. 2008. Evolutionary transitions in floral color. International Journal of Plant Science 169: 7–21.

[113] What plant evo-devo is, where it came from and where it's going is considered in Friedman WE, Barrett SCH, Diggle PK, Irish VF, Hufford L. 2008. Whither plant evo-devo? New Phytologist 178: 468–472.

[114] Species of the genus *Agave* give us aloe, sisal and tequila, among other useful products. *Agave* spp. are slow-growing monocarpic perennials that form rosettes of large, thick, leathery leaves until they switch from vegetative to reproductive development. Whereupon they produce tall flowering stems ('masts') and die after fruiting. The extended vegetative period followed by monocarpic death has led to the name 'century plant' for some species, though the lifespan is usually no more than 20 to 40 years. As pandas experience to their cost, flowering and monocarpic death are synchronised in long-lived woody bamboos across the continents (for example, Franklin DC. 2004. Synchrony and asynchrony: observations and hypotheses for the flowering wave in a long-lived semelparous bamboo. Journal of Biogeography 31: 773–786). A spectacular recently discovered example of monocarpic death in a long-lived perennial is the rare so called 'suicide palm' of Madagascar, *Tahina spectabilis* (Dransfield, J., Rakotoarinivo, M., Baker, W. J., Bayton, R. P., Fisher, J. B., Horn, J. W., Leroy, B. And Metz, X. (2008), A new Coryphoid palm genus from Madagascar. Botanical Journal of the Linnean Society 156: 79–91). For images of this endangered species, see http://www.kew.org/science-conservation/plants-fungi/tahina-spectabilis-dimaka [accessed 23 February 2016].

[115] The functional ecologists (see note 91) represent another tribe with which there ought to be more contact, but who generally stick to their side of what amounts to a cultural divide. I feel certain that a rainbow coalition forged from the diversity of research perspectives would greatly enrich our understanding of senescence and ageing.

[116] William Donald Hamilton (1936-2000). The quotation is from Hamilton WD. 1966. The moulding of senescence by natural selection. Theoretical Biology 12: 12-45.

[117] Skulachev VP. 2011. Aging as a particular case of phenoptosis, the programmed death of an organism (A response to Kirkwood and Melov "On the programmed/nonprogrammed nature of ageing within the life history"). Aging 3: 1120-1123.

[118] Humphries S, Stevens DJ. 2001. Reproductive biology: out with a bang. Nature 410: 758-759.

[119] See Thomas (2013) – note 4.

[120] All about darnel: see Thomas H, Archer J, Marggraf Turley R. 2011. Evolution, physiology and phytochemistry of the psychotoxic arable mimic weed darnel (*Lolium temulentum* L). Progress in Botany 72: 73-104; Thomas H, Archer JE, Marggraf Turley R.

2016. Remembering darnel, a forgotten plant of literary, religious and evolutionary significance. Journal of Ethnobiology 36: 29-44.

[121] See: Thomas H, Thomas HM, Ougham H. 2000. Annuality, perenniality and cell death. Journal of Experimental Botany 51: 1781-1788; and Thomas (2013) – note 4.

[122] Czech-Austrian botanist, born 1856, died 1937. He held professorships in Prague, Vienna, Sendai (Japan) and India.

[123] For the concept of senescence as the price paid for sex, see: Stearns SC. 1989. Trade-offs in life-history evolution. Functional Ecology 3: 259-268.

[124] Prevention of soybean senescence by removing flowers was first reported by Leopold AC, Niedergang-Kamien E, Janick J. 1959. Experimental modification of plant senescence. Plant Physiology 34: 570-573. Subsequently the same response was observed when developing pods or seeds were removed (Lindoo SJ, Noodén LD. 1977. Studies on the behavior of the senescence signal in Anoka soybeans. Plant Physiology 59: 1136-1140). In lectures, Larry Noodén often showed a picture of a repeatedly de-podded soybean that filled a greenhouse. I haven't been able to find a copy, but believe me, it was spectacular.

[125] Wilson JB. 1997. An evolutionary perspective on the 'death hormone' hypothesis in plants. Physiologia Plantarum 99: 511-516.

[126] See note 10

[127] See Leopold et al. (1959) – note 124.

[128] See, for example, Bonduriansky R, Maklakov A, Zajitschek F, Brooks R. 2008. The evolutionary ecology of senescence: sexual selection, sexual conflict and the evolution of ageing and life span. Functional Ecology 22: 443–453.

[129] The painting is from a website that seems to have gone offline. If you are the artist, please contact to discuss permission to use your beautiful work.

[130] A database of ancient trees can be found here: www.rmtrr.org/oldlist.htm [accessed 23 February 2016].

[131] Lynch AJJ, Barnes RW, Cambecèdes J, Vaillancourt RE. 1998. Genetic evidence that *Lomatia tasmanica* (Proteaceae) is an ancient clone. Australian Journal of Botany 46: 25–33.

[132] Ally D, Ritland K, Otto SP. 2010. Aging in a long-lived clonal tree. PLoS Biology 8(8): e1000454 (doi:10.1371/journal.pbio.1000454).

[133] See Lanner RM, Connor KF. 2001. Does bristlecone pine senesce? Experimental Gerontology 36: 675-685; and Ally et al. (2010) – note 132.

[134] Born 1856, died 1950. The quotation is from *Man and Superman* (1903).

[135] Vaupel JW, Baudisch A, Dölling M, Roach DA, Gampe J. 2004. The case for negative senescence Theoretical Population Biology 65: 339–351.

[136] Here are some informative publications about the size and structure of trees in relation to ageing and lifespan: Day ME, Greenwood MS, Diaz-Sala C. 2002. Age- and size-related trends in woody plant shoot development: regulatory pathways and evidence for genetic control. Tree Physiology 22: 507–513; Johnson SE, Abrams MD . 2009. Age class, longevity and growth rate relationships: protracted growth increases in old trees in the eastern United States. Tree Physiology 29: 1317–1328; Ryan MG, Yoder BJ. 1997. Hydraulic limits to tree height and tree growth. BioScience 47: 235-242; Hubbard RM, Bond BJ, Ryan MG. 1999. Evidence that hydraulic conductance limits photosynthesis in old *Pinus ponderosa* trees. Tree Physiology 19: 165-172; Ishii HT, Ford ED, Kennedy MC. 2007. Physiological and ecological implications of adaptive reiteration as a mechanism for crown maintenance and longevity. Tree Physiology 27: 455–462. Lanner and Connor (2001) – see note 133 – found no evidence for size-related symptoms of ageing in bristlecone pine.

[137] Alfred Joyce Kilmer, American poet, born 1886, killed by a sniper at the Second Battle of the Marne, 1918.

[138] Katharine Elizabeth Whitehorn (born 1928), journalist and author. The quotation is from *Observations* (Methuen, 1970).

[139] For a contemporary view of the shoot apex as the centre of a developmental network, see Traas J, Monéger F. 2010. Systems biology of organ initiation at the shoot apex. Plant Physiology 152: 420-427.

[140] The suppression of axillary bud growth by the shoot apex is called apical dominance, a classic subject of plant physiological research. The growth of lateral branches is inhibited by hormones produced by the apical meristem. This article explains: Leyser O. 2005. The fall and rise of apical dominance. Current Opinion in Genetics and Development 15: 468–471.

[141] If you keep your eyes open, especially in late summer and early autumn, you can find many very beautiful examples like this of the war between plant and pathogen. The original photograph was taken by my old boss, John L Stoddart.

[142] There are online animations of the different patterns of growth and senescence linked to the review by Thomas (2013) – note 4.

[143] The formation of spreading networks of ramets allows clonal plants to exploit patchy habitats. The term 'foraging', more usually applied to animal behaviour, was introduced by Sutherland WJ, Stillman RA. 1988. The foraging tactics of plants. Oikos 52: 239–244.

[144] The concept of symbiosis was introduced by Heinrich Anton de Bary (1831-1888), one of the founders of modern mycology and phytopathology: de Bary HA. 1879. Die Erschenung Symbiose. In: Trubner KJ, editor. *Vortrag auf der Versammlung der Naturforscher und Artze zu Cassel* (Strassburg) pp. 1-30.

[145] The term 'allelopathy' was coined by Hans Molisch (see note 122) and refers to the beneficial or harmful effects of one plant on another, exerted through the release (usually from the roots) of bioactive chemicals.

[146] For an accessible general introduction to the fungi and fungus-like microbes, including pathological and symbiotic interactions with plants and other organisms, see Deacon J. 2006. *Fungal Biology* (Blackwell). Useful overviews of bacterial, nematode and viral diseases of plants are given in the following publications of the American Phytopathological Society (online at www.apsnet.org/edcenter/intropp/PathogenGroups/Pages/default.aspx [accessed 23 February 2016]): Vidaver AK, Lambrecht PA. 2004. Bacteria as plant pathogens (doi:10.1094/PHI-I-2004-0809-01); Lambert K, Bekal S. 2002, revised 2009. Introduction to Plant-Parasitic Nematodes (doi:10.1094/PHI-I-2002-1218-01); and Gergerich RC, Dolja VV. 2006. Introduction to Plant Viruses, the Invisible Foe (doi:10.1094/PHI-I-2006-0414-01).

[147] The DNA sequence of the *P. infestans* genome was published by an international consortium in 2009 (Haas BJ et al. 2009. Genome sequence and analysis of the Irish potato famine pathogen *Phytophthora infestans* Nature 461: 393-398) and has given new insights into the genetic basis of the organism's adaptability and virulence.

[148] Shakespeare was fond of allusions to 'blast' – for example, in the 'blasted heath' of *Macbeth*. It is often supposed to refer to ravage by lightning, but this is wrong. Shakespeare was almost certainly revelling in the opportunity to employ a newly minted word– the OED records that the earliest recorded use of 'blast' in the sense of the fungal disease occurs in Barnabe Gooch's 1577 translation of Conrad Heresbach's *Foure Bookes of Husbandry*. The error has been further compounded in modern performances of *King Lear*, where the eponymous senescent monarch rails madly against the storm. Not only should the scene be set in a landscape subject to another meaning of blast, but the direction 'blasted heath' does not even occur in the Folio, being an 18th century amendment to the text (probably recycled from *Macbeth*). For a thorough examination of these matters, see JE Archer, H Thomas, R Marggraf Turley. 2012. The Autumn King: remembering the land in King Lear. Shakespeare Quarterly 63: 518-543, and Archer et al. 2014. *Food and the Literary Imagination* – note 50.

[149] *Pseudomonas syringae* secretes harpin, a protein that elicits pathogenic or hypersensitive reactions in host or non-host plants respectively. *P. syringae* has another way of tormenting

plants too: it secretes Ice Nucleation Protein, which facilitates the formation of ice crystals in host cells, resulting in frost damage. See: He S, Yang S, Huang H, Collmer A. 1993. *Pseudomonas syringae* pv. syringae harpin$_{Pss}$: a protein that is secreted via the hrp pathway and elicits the hypersensitive response in plants. Cell 73: 1255-1266; Li Q, Yan Q, Chen J, He Y, Wang J, Zhang H, Yu Z, Li L. 2012. Molecular characterization of an Ice Nucleation Protein variant (InaQ) from *Pseudomonas syringae* and the analysis of its transmembrane transport activity in *Escherichia coli*. International Journal of Biological Sciences 8: 1097-1108.

[150] The continuum of cells physically connected by plasmodesmata constitutes the symplast, a network across which small molecules can travel freely, and larger structures such as proteins and viruses can be moved on subcellular conveyor systems. The space outside the symplast is called the apoplast. For more on this, see *The Molecular Life of Plants* (note 9).

[151] The genetic variants of Ug99 are being tracked (and sobering reading it makes): see the CIMMYT Rust Pathotype Tracker http://rusttracker.cimmyt.org/?page_id=22 [accessed 23 February 2016]. CIMMYT (the International Maize and Wheat Improvement Center) posted an alert for the Eastern Mediterranean in April 2015 following identification of a particularly virulent race of Ug99 in Egypt.

[152] The array of chemical weapons that plants have evolved has always been a major source of drugs employed for medical (and recreational) purposes. Over a hundred pharmaceuticals and other bioactives in everyday use are of plant origin, including (to name only a few beginning with 'c') caffeine, camphor, camptothecin, cocaine, codeine, and colchicine. The central importance of plant chemicals in drug discovery and design is addressed in Cechinel-Filho V (editor). 2012. *Plant Bioactives and Drug Discovery: Principles, Practice, and Perspectives* (Wiley).

[153] Elvin Charles Stakman (1885-1979), plant pathologist, mentor of Norman Borlaug (see note 49) and implacable foe of 'these shifty little enemies that destroy our food crops'.

[154] See Mur LA, Kenton P, Lloyd AJ, Ougham H, Prats E. 2008. The hypersensitive response; the centenary is upon us but how much do we know? Journal of Experimental Botany 59: 501-520.

[155] It takes a lot of energy to split a molecule of water into hydrogen and oxygen (around 287 kiloJoules, corresponding to an electrical voltage of 1.48V). Photosynthesis achieves this by capturing light energy and intensifying it in the reaction centres and water-splitting complexes of the internal membranes of the chloroplast. As a consequence, oxygen production during photosynthesis is a source of highly reactive chemical species (ROS) that must be contained or quenched if damage to cells is to be avoided. ROS are produced during many metabolic processes and are implicated in a range of stress responses, regulatory pathways and pathological changes in living organisms (Murphy MP, Holmgren A, Larsson NG, Halliwell B, Chang CJ, Kalyanaraman B, Rhee SG, Thornalley PJ,

Partridge L, Gems D, Nyström T, Belousov V, Schumacker PT, Winterbourn CC. 2011. Unraveling the biological roles of reactive oxygen species. Cell Metabolism 13: 361-366)

[156] *The Emerald Planet: How plants changed Earth's history* (OUP, 2007). This fine book was the inspiration for a rather good BBC television series, *How to Grow a Planet*, engagingly presented by geologist Iain Stewart.

[157] A nice quotation from a not very nice man. He was an influential philosopher, whose reputation is now that of a precocious near-genius disfigured by misogyny and anti-semitism (though he was of Jewish extraction himself). In 1903, at the age of 23, he committed suicide in the hotel room where Beethoven died.

[158] Also, of course, the agent responsible for 'peroxide blonde' hair. The oxygen released from hydrogen peroxide reacts with the melanin that gives hair its dark colour and bleaches it. The 'platinum blonde' look that made the actress Jean Harlow famous was reputedly achieved with a mixture of peroxide, ammonia, Clorox and Lux flakes. Little wonder Ms Harlow's hair fell out and she was dead by the age of 26.

[159] A molecule becomes oxidised when oxygen is added to it or, more generally, when it loses an electron. The acceptor of the electron from the oxidised molecule is said to have become reduced. A redox reaction is the term for a chemical change in which one participant is the electron donor and becomes oxidised and another is the electron acceptor and becomes reduced. Metabolism, the biochemistry of living cells, largely consists of redox reactions that serially move electrons between donors and acceptors. A flow of electrons is, by definition, an electric current. It's easy to see where Victor Frankenstein got the idea of rejuvenation through the power of lightning (see note 79).

[160] The free radical theory, which considers biological ageing to be the consequence of the progressive accumulation over time of cell damage caused by free radicals, was first proposed by the gerontologist Denham Harman (1916-2014) – see Harman D. 1956. Aging: a theory based on free radical and radiation chemistry. Journal of Gerontology 11: 298–300. This has given rise to the promotion of antioxidants of every kind as anti-ageing measures, either applied to the outside (in shampoos, cosmetics and unguents) or ingested as foodstuffs, medicines and nutritional supplements. Experimental evidence in support of the efficacy of antioxidants in the treatment or prevention of ageing and age-related disorders is at best ambiguous. Harman's theory has also influenced a whole research industry devoted to linking plant senescence with free radicals – see, for example, Leshem YY, Halevy AH, Frenkel C (editors). 2012. *Processes and Control of Plant Senescence: Developments in Crop Science* (Elsevier). I remain a sceptic.

[161] Particularly catalase, superoxide dismutase, peroxidases, dehydroascorbate reductase and glutathione reductase. The interactions and cycles in which these enzymes and ROS participate are networks of redox reactions (see note 159). Enzymic and non-enzymic processes that detoxify ROS are reviewed in Apel K, Hirt H. 2004. Reactive oxygen species:

metabolism, oxidative stress, and signal transduction. Annual Review of Plant Biology 55: 373-399.

[162] See Apel and Hirt (2004).

[163] Sidney Joseph Perelman (1904-1979), American humourist, often (to his intense annoyance) associated with the Marx Brothers on the strength of contributions to the screenplays of *Monkey Business* (1931) and *Horse Feathers* (1932).

[164] See notes 97 and 98.

[165] See, for example, Mur LAJ, Aubry S, Mondhe M, A Kingston-Smith A, Gallagher J, Timms-Taravella E, James C, Papp I, Hörtensteiner S, Thomas H, Ougham H. 2010. Accumulation of chlorophyll catabolites photosensitizes the hypersensitive response elicited by *Pseudomonas syringae* in Arabidopsis. New Phytologist 188: 161–174.

[166] This may at first seem like an irrelevant diversion, but bear with me. The Green Revolution was aimed largely at solving the primary dilemma of global malnutrition, that of insufficient calories in the diet. Once the calorie threshold has been crossed, however, the next problem facing world food supply is that of insufficient protein. The protein in grains and pulses is substantially derived from amino acids remobilised from the proteins of senescing leaves. The question therefore arises: why not cut out the inefficiencies of nitrogen transfer from vegetative tissues to seeds and use foliage as a source of proteins for the human diet. Of course people can't live by grazing on grass leaves, but there are various simple technologies for extracting and concentrating leaf protein. Prominent among the advocates for leaf protein was Norman Wingate Pirie (1907-1997), a distinguished scientist who, before turning to philanthropically motivated nutritional research at Rothamsted Experimental Station, had determined the structure of tobacco mosaic virus. The story of leaf protein is told in Pirie NW. 1987. *Leaf Protein and its By-products in Human and Animal Nutrition* (Cambridge University Press). One problem with the extraction procedure is that it produces protein that is green in colour, because it adsorbs products of chlorophyll degradation. Not only do they have a strong grassy taste, leaf protein preparations are visually unappealing and people tend not to find them very appetising. Moreover, the chlorophyll derivatives provoke photosensitization symptoms in some people. Margaret Holden, Pirie's colleague who had the job of solving the chlorophyll problem, used to visit my lab now and then in search of new ideas (I wasn't able to be very helpful, sadly). Ultimately the leaf protein project faltered, but I can't help wondering whether modern technology might not easily overcome the difficulties. Some interest in leaves as protein sources continues in the United States and Central and South America. See: www.leafforlife.org/index.htm [accessed 3 May 2015].

[167] Reviewed in Bruggeman Q, Raynaud C, Benhamed M, Delarue M. 2015. To die or not to die? Lessons from lesion mimic mutants. Frontiers in Plant Science 6: 24 (doi:10.3389/fpls.2015.00024).

[168] The relation between the Sekiguchi lesion character and rice blast resistance was first reported by Kiyosawa, S. 1970. Inheritance of a particular sensitivity of the rice variety, Sekiguchi Asahi, to pathogens and chemicals, and linkage relationship with blast resistance genes. Nogyo Gijutsu Kenkyusho Hokoku (Bulletin of the National Institute of Agricultural Science) 21: 61-72. For the first description of *mlo* and barley mildew resistance, see Kjaer B, Jensen HP, Jensen J, Jorgensen JH. 1990. Associations between three *mlo* powdery mildew resistance genes and agronomic traits in barley. Euphytica 46: 185 193.

[169] The plagues of Biblical times remain with us. The Food and Agriculture Organisation (FAO) of the United Nations has recognised a state of crisis in Madagascar, which has been subject to successive outbreaks of locust infestations since 2012, threatening the livelihoods of 13 million people. One million locusts can eat about one tonne of food each day, and the largest swarms can consume over 100,000 tonnes each day, or enough to feed tens of thousands of people for one year. The FAO emergencies website is www.fao.org/emergencies/emergency-types/plant-pests-and-diseases/en/ [accessed 23 February 2016].

[170] The precision with which an aphid can introduce its stylet (feeding tube) into the conductive tissue (phloem) of the plant host has been exploited by physiologists to obtain samples of sap for chemical analysis. When the stylet of a feeding insect is severed (using very fine scissors or a laser), a drop of phloem contents is exuded. Strictly speaking, therefore, aphids are not sap-suckers but rather are passive beneficiaries of the pressure within the phloem. The aphid sampling technique was originally developed by Kennedy JS, Mittler TE. 1953. A method of obtaining phloem sap via the mouth-parts of aphids. Nature 171: 528.

[171] The standard work is the three volume *Plant Parasitic Nematodes* (Elsevier, 2012), edited by Bert Zuckerman. Studies of plant nematodes have greatly benefitted from the scale of research effort and resources devoted to *C. elegans* as a related model organism – see note 74.

[172] Volatile organic compounds have multiple roles in addition to defence against biotic and abiotic stresses. They attract pollinators and seed dispersers, and function in plant to plant signalling. See Dudareva N, Klempien A, Muhlemann JK Kaplan I. 2013. Biosynthesis, function and metabolic engineering of plant volatile organic compounds. New Phytologist 198: 16–32.

[173] This kind of shoot apical meristem is termed intercalary. It's typical of monocots, that is, grasses and other families such as the orchids, palms, lilies and amaryllids. If you peel away the layers of an onion (a member of a sub-family of the amaryllids) you will find, right at the heart, the intercalary meristem. That's what a bulb is – an intercalary meristem surrounded by layers of swollen leaf bases.

[174] John Raven and I have written a brief evolutionary history of grasses and grazers – see Raven J, Thomas H. 2010. Quick guide: grasses. Current Biology 20: R837-R839.

[175] See Kingston-Smith AH, Theodorou MK. 2000. Post-ingestion metabolism of fresh forage. New Phytologist 148: 37-55; Kingston-Smith AH, Davies TE, Edwards JE, Theodorou MK. 2008. From plants to animals; the role of plant cell death in ruminant herbivores. Journal of Experimental Botany 59: 521-532.

[176] The background to the fate of ingested forage in the rumen is discussed by Huws SA, Mayorga OL, Theodorou MK, Kim EJ, Cookson AH, Newbold CJ, Kingston-Smith AH. 2014. Differential colonization of plant parts by the rumen microbiota is likely to be due to different forage chemistries. Journal of Microbial and Biochemical Technology 6: 80-86.

[177] These thoughts were first expressed in a forgotten publication from long ago: Thomas H. 1998. Air today - gone tomorrow. New Phytologist 139: 225-229. I'm quite fond of this paper, even though it has only ever been cited three times, because (a) it was dedicated to my father, and (b) it contains the (insightful, in my opinion) statement 'poor countries smell of faeces, affluent countries of urine'.

[178] Senescence played a part in the origin of the cell concept of biological organisation. Robert Hooke (1635-1703) is generally credited with introducing the term 'cell', in his pioneering account of microscopy, *Micrographia: or Some Physiological Descriptions of Miniature Bodies Made by Magnifying Glasses* (1665). On examining a thin slice of cork, he observed empty spaces contained by walls, and called them pores or cells. Mature cork cells are empty as a result of programmed senescence and death of cytoplasm during the development of cork tissue (phellem). For a general account of cork and its uses, see Pereira H. (ed). 2011. *Cork: Biology, Production and Uses* (Elsevier). Recent studies of cell death during cork development include: Ricardo CPP, Martins I, Francisco R, Sergeant K, Pinheiro C, Campos A, Renaut J, Fevereiro P. 2011. Proteins associated with cork formation in *Quercus suber* L. stem tissues. Journal of Proteomics 74: 1266-1278; and Chaves I, Lin YC, Pinto-Ricardo C, Van de Peer Y, Miguel C. 2014. miRNA profiling in leaf and cork tissues of *Quercus suber* reveals novel miRNAs and tissue-specific expression patterns. Tree Genetics & Genomes 10: 721-737. The illustration is of a generalised green plant cell, showing nucleus (purple), central vacuole (blue) and chloroplasts (green), as well as mitochondria, endoplasmic reticulum, Golgi bodies and cytoskeleton structures.

[179] The quotation is from Archetti M. 2009. Classification of hypotheses on the evolution of autumn colours. Oikos 118: 328-333. Marco has been a leader in bringing together the various tribes (see notes 91, 115) to address the question of the biological meaning of colour change during senescence.

[180] See Archetti M, Döring TF, Hagen SB, Hughes NM, Leather SR, Lee DW, Lev-Yadun S, Manetas Y, Ougham HJ, Schaberg PG, Thomas H. 2009. Unravelling the evolution of

autumn colours - an interdisciplinary approach. Trends in Ecology and Evolution 24: 166-173.

[181] A bilin (see note 188), informally referred to as RCC (Red Chlorophyll Catabolite).

[182] The quotation is from *The Queen of the Air: Being a Study of the Greek Myths of Cloud and Storm* (Library of Alexandria, 1869).

[183] 'Why grass is green, or why our blood is red, Are mysteries which none have reach'd unto' wrote John Donne in 1612, sensing perhaps a commonality. Chlorophyll and haem (heme in US English) share a basic molecular structure of carbon and nitrogen atoms arranged in a square with a pentagon at each corner. Each pentagon consists of one nitrogen and four carbon atoms and is referred to as a pyrrole group – and so the basic structural unit of chlorophyll and haem is called a tetrapyrrole. Tetrapyrroles arose at the dawn of life on Earth and so versatile were they that they quickly evolved into structural units for a diversity of functions in living cells. The configuration of a square with pentagons at each corner forms a kind of nest that can hold a metal atom. In chlorophyll, the metal is magnesium. In haem, it's iron. The enzyme that fixes atmospheric nitrogen, nitrogenase, has a tetrapyrrole with a central molybdenum atom. And the chemical structure of vitamin B_{12} is based on a tetrapyrrole with a cobalt atom.

[184] The standard work on chlorophylls is Scheer H. 1991. *Chlorophylls* (CRC). Recent developments in chlorophyll chemistry and physiology are covered by Grimm B (editor). 2007. *Chlorophylls and Bacteriochlorophylls: Biochemistry, Biophysics, Functions and Applications* (Springer).

[185] Avram Noam Chomsky (born 1928), linguistician (is that a word? He would know) and social activist. The quotation is his famous example (from *Syntactic Structures*, 1957) of a meaningless but syntactically correct sentence. Except, for those of us in the chlorophyll catabolism business, it's meaningful enough.

[186] Anabolism is the term for synthesis corresponding to catabolism for breakdown. Metabolism is the general name for biochemical changes in living cells. Metabolic turnover is the sum of anabolism and catabolism. Jargon, jargon.

[187] Philippe Matile (1931-2011), cell biologist. Phibus (as he was universally known) was a brilliant, cultured, amusing, talented, unique man who made important contributions to our understanding of many aspects of botany, including senescence.

[188] Cutting one side of the tetrapyrrole square (see note 183) results in a linear tetrapyrrole. Such molecular structures are generally called bilins, named after bile, the brownish-green colour of which is due to the products of haem catabolism that are eliminated from the body by this route. Chlorophyll catabolites are bilins, as are the photoreactive centre of phytochrome (see note 55) and the photosynthetic pigments of blue-green and red algae.

[189] Much of this research has been led by my friend Stefan Hörtensteiner in Zürich, carrying on the Matile tradition (note 187). Recent advances are reviewed in Christ B, Hörtensteiner S. 2014. Mechanism and significance of chlorophyll breakdown. Journal of Plant Growth Regulation 33: 4-20.

[190] The first murder on a train took place in 1864 on the 09:50 North London Railway service from Fenchurch Street. The victim was a banker, Thomas Briggs. The murderer, Franz Muller, attempted to escape by ship but was arrested in New York, extradited, convicted and executed. The case led directly to the installation of communication chords in compartments, but carriages continued to be corridor-less for decades thereafter. From the earliest days of the railways, carriages have been the scenes and means of crimes (train journeys are as evocative as London fog in the exploits of Sherlock Holmes) and their sinister associations continue into the modern era. W.H. Auden's terrifying 'Gare du Midi' (1938) catches the mood all too perfectly: 'A nondescript express in from the South...Clutching a little case/He walks out briskly to infect a city/Whose terrible future may have just arrived.'

[191] Plastids are organelles unique to plants. They, like mitochondria, are thought to have originated in early evolution as free-living bacteria-like cells that were captured by the host cell, with which they settled into a symbiotic relationship. Plastids and mitochondria retain their own genomes that are clearly of bacterial origin, although over evolutionary time many of their genes have moved to the nuclear genome or have been lost altogether. Plastids form a developmental network of different kinds of organelle and have the capacity to change from one form to another. As well as the chloroplast, the green, photosynthetic plastid, there is the amyloplast (a colourless plastid that stores starch in, for example, potato tubers), the proplastid (the undifferentiated organelle from immature cells), the etioplast (the incipient chloroplast of tissues kept in the dark) and the chromoplast (the brightly coloured plastid of fruits and flowers). The plastid of senescing leaves has been called the gerontoplast (Parthier B. 1988. Gerontoplasts: the yellow end in the ontogenesis of chloroplasts. Endocytobiosis and Cell Research 5: 163-190). This is a handy term, serving to integrate this stage of development into the network of plastid differentiation, but even though I use it freely myself, I'm a bit uneasy about the suggested connection with gerontology. I have played with other, more appropriate, terms based on classical Greek roots. Thus athroisoplast, from the Greek for mobilisation, *athroisis*. Or perigenomenoplast, from *perigenomena*, salvaged things. Maybe apolambaneoplast, from *apolambanein*, reclaim. None of these is remotely practical, of course. But how about karposoplast, from *karpos*, harvest? If we could rewind the clock, I would put in a strong recommendation that karposoplast, rather than gerontoplast, should be the correct term. The faint scraping sound in the background is John Ruskin turning in his grave.

[192] A more or less inevitable quotation. Dylan Thomas was born in Swansea in 1914 and died (of 'everything') in New York City, 1953. My friend and occasional musician colleague Daniel Williams likes to call him the Welsh Charlie Parker – see Williams DG. 2015. *Wales*

Unchained: Literature, Politics and Identity in the American Century (University of Wales Press).

[193] 'Your molecular structure' is a 1968 song by 'The Sage of Tippo', Mose John Allison (born 1927), influential composer and performer with a unique and irreproducible piano style someone once perceptively described as 'romping'.

[194] The complex membrane systems within the chloroplast are called thylakoids and are the sites of light capture, biological energy generation, and oxygen production from the splitting of water.

[195] The light-harvesting chlorophyll-protein complexes are closely associated with carotenoids, accessory pigments that protect against the hazards of ROS and free radical propagation.

[196] A protein is a string of (up to several hundred) amino acids joined end-to-end. There are 20 chemically different amino acids, and the number of copies of each type of amino acid, and the order in which they are arranged along the protein string, determines the properties of the protein. Protein structure is in turn specified by the corresponding gene, a relationship at the heart of the DNA revolution started by Watson and Crick. Proteases are enzymes that break up proteins by snipping the bonds between neighbouring amino acids. The proteases of pineapple (bromelain) and papaya (papain) are responsible for the meat-tenderising qualities of these fruits. *The Molecular Life of Plants* (see note 9) explains all.

[197] One per pyrrole group – see note 183.

[198] Rubisco is short for ribulose-1,5-bisphosphate carboxylase-oxygenase. The term is said to have originated as a humorous coinage by the protein chemist David Eisenberg.

[199] See Raven JA. 2013. Rubisco: still the most abundant protein of Earth? New Phytologist 198: 1–3.

[200] Chlorophyll and its associated proteins are embedded in the thylakoid membranes, whereas rubisco is located in the stroma, the soluble matrix of the chloroplast.

[201] The emerging view is that dismantling the chloroplast can take place by any of at least three pathways. One of these, autophagy, is an evolutionarily conserved process that has been studied in molecular detail. The other two (the senescence-associated vacuole system and the stress-induced plastid vesiculation route) are plant-specific and not so well understood, but they attract increasing research attention. See Xie Q, Michaeli S, Peled-Zehavi H, Galili G. 2015. Chloroplast degradation: one organelle, multiple degradation pathways. Trends in Plant Science 20: 264–265.

[202] See note 187.

[203] For a personal view of the history of lysosomes by their discoverer (and coiner of the term 'autophagy'), Nobel laureate Christian René Marie Joseph, Viscount de Duve (1917–2013), see Klionsky, DJ. 2008. Autophagy revisited: A conversation with Christian de Duve. Autophagy 4: 740–743.

[204] It famously features in the 1953 motion picture *Roman Holiday*, starring Gregory Peck and Audrey Hepburn. Coin-operated mechanical versions of the Bocca are widely distributed. There used to be one at Magor Services on the M4 motorway in South Wales.

[205] The quotation is from the short story 'Love will tear us apart', in Black H, Larbalestier J (editors). 2012. *Zombies vs. Unicorns* (Margaret K. McElderry Books).

[206] Unexpectedly, autophagy is turning out to have a material influence on the global carbon cycle, and thereby to be a player in climate change. *Emiliana huxleyi* a single-celled marine planktonic alga, is a very beautiful organism when seen under the microscope. It covers itself with cartwheel-shaped scales of calcium carbonate. *E. huxleyi* forms enormous blooms, visible in images from earth observation satellites, and by capturing huge quantities of atmospheric CO_2 and locking it up in their scales (which accumulate over geological time as limestone), these population explosions are critical to maintaining the balance of greenhouse gases. The boom in *E. huxleyi* numbers is inevitably followed by a bust, largely as the consequence of an epidemic of infection by a specific (and unusually large) virus. The virus takes over the host's genetic system to replicate itself, and then triggers autophagy as part of the assembly and release process. See Schatz D, Shemi A, Rosenwasser S, Sabanay H, Wolf SG, Ben-Dor S, Vardi A. 2014. Hijacking of an autophagy-like process is critical for the life cycle of a DNA virus infecting oceanic algal blooms. New Phytologist 204: 854-863.

[207] Current thinking about autophagy as the mechanism for chloroplast protein mobilisation is reviewed in Avila-Ospina L, Moison M, Yoshimoto K, Masclaux-Daubresse C. 2014. Autophagy, plant senescence, and nutrient recycling. Journal of Experimental Botany 65: 3799-3811. These authors point out a continuing area of ignorance: there is little evidence that the cargo of rubisco and other non-membrane proteins carried by the vesicles plying their trade between chloroplast and vacuole also includes thylakoid proteins, such as those of the chlorophyll-binding complexes. We are left to conclude that there must be another salvage system for plastid membrane proteins, separate from the vesicular route that deals with rubisco and other stroma constituents. There may yet be a role for proteases located, not in the vacuole but in the chloroplast itself – see Roberts IN, Caputo C, Criado MV, Funk C. 2012. Senescence-associated proteases in plants. Physiologia Plantarum 145: 130–139. But we are far from getting the full story of the fate of chloroplast proteins in senescence.

[208] The labyrinth-like membrane system of the cytoplasm is the endoplasmic reticulum, another of the cell's compartments, the contents of which are physically separated from the

constituents of other compartments. Our limited knowledge of the nature, subcellular locations and functions in senescence of the DNA-degrading enzymes of plants have been reviewed by Sakamoto W, Takami T. 2014. Nucleases in higher plants and their possible involvement in DNA degradation during leaf senescence. Journal of Experimental Botany 65: 3835-3843.

[209] The picture is a detail from *Gregor Mendel counting green and yellow peas in the Augustinian Abbey at Brno*. It was painted in the Early Renaissance style by Phibus Matile (see note 187). You may notice a little detail that refers to what has become known as the Mendel-Fisher controversy. The terrifyingly eminent statistician, Sir Ronald Aylmer Fisher (1890–1962), re-analysed Mendel's inheritance experiments and bluntly asserted, that 'the data of most, if not all, of the experiments have been falsified so as to agree closely with Mendel's expectations' (Fisher RA. 1936. Has Mendel's work been rediscovered? Annals of Science 1: 115-137). There are two ways of looking at this. One is that Mendel, being a genius, had worked out the laws of inheritance by sheer brainpower and subsequently made sure the numbers supported his theory, as Fisher stated. The other is that he is innocent of manipulating his data and that the statistical anomalies can be explained without impugning the reputation of someone who was, after all, a man of the cloth. It is pretty well accepted now that the latter is the true explanation (see Hartl DL, Fairbanks DJ. 2007. Mud sticks: on the alleged falsification of Mendel's data. Genetics 175: 975-979). In Matile's painting, Mendel is counting green and yellow peas, unaware that some yellow peas have fallen to the floor beneath the table. That's how easy it is for data to deceive. You may also note that the pea plants being tended in the garden are yellow and green in the ratio 3:1. This picture, which hangs on my wall, is one of my most treasured possessions.

[210] I eventually got round to publishing the inheritance study: Thomas H. 1987. *Sid*: a Mendelian locus controlling thylakoid membrane disassembly in senescing leaves of *Festuca pratensis*. Theoretical and Applied Genetics 73: 551–555. I should perhaps explain the reason for the presence of *Sid* in the title. After I had drafted the paper, I asked people in the lab to suggest a suitable name for the gene, bearing in mind the informal convention that it should ideally be a three-letter abbreviation or acronym. They came back with *Sid* and of course I immediately rejected it on the grounds of modesty (*Sid* is the soubriquet by which I'm known by friends and colleagues). Then I thought, well, it might be my one chance of immortality, so I overcame my natural diffidence and, in a burst of self-indulgence (to which the present book shows I am sometimes inclined), I went ahead with it. Now the gene is called *SGR*, and my bid for immortality has failed. There's a moral here somewhere.

[211] The quotation is from *Peas in our time*, a song presented by my friend, colleague and sometime musical collaborator Wendy Silk in the 'Song of Botany' session at the XVIII International Botanical Congress, Melbourne, Australia (2011). Helen Ougham wrote the lyrics in the time-honoured fashion, on a table napkin in a restaurant, and I subsequently set them to music.

[212] The progeny of a cross between green and yellow lines of pea are all yellow. In other words, yellow is the dominant trait, green the recessive. It's interesting how often students get this wrong and think that the offspring must be green. Presumably the assumption is that mutations are generally recessive and that turning yellow looks like the abnormal behaviour of a mutant whereas green looks healthy. The dominance of yellowing is the clearest evidence that senescence is a normal, genetically regulated phase of leaf development.

[213] Thomas H, Schellenberg M, Vicentini F, Matile P. 1996. Gregor Mendel's green and yellow pea seeds. Botanica Acta 109: 3-4. Botanica Acta was the successor to Berichte der Deutschen Botanischen Gesellschaft, the journal of the German Botanical Society. We chose to submit our paper to this not particularly high-profile journal (in defiance of the imperative to aim for the highest possible impact at all costs) because the Berichte was where Carl Erich Correns (1864-1933) published the rediscovery of Mendel's work (Correns C. 1900. G. Mendel's Regel über das Verhalten der Nachkommenschaft der Rassenbastarde. Berichte der Deutschen Botanischen Gesellschaft 18: 158–168).

[214] Some reviews of stay-green: Thomas H, Smart CM. 1993. Crops that stay green. Annals of Applied Biology 123: 193-219; Thomas H, Howarth CJ. 2000. Five ways to stay green. Journal of Experimental Botany 51: 329-337; see also Thomas and Ougham (2014; note 63).

[215] Thomas and Ougham (2014; note 63) includes a detailed discussion of the cell biology and physiology of cosmetic stay-greens.

[216] Armstead I, Donnison I, Aubry S, Harper J, Hörtensteiner S, James C, Mani J, Moffet M, Ougham H, Roberts L, Thomas A, Weeden N, Thomas H, King I. 2006. From crop to model to crop: identifying the genetic basis of the staygreen mutation in the forage grass *Festuca pratensis* (Huds.) New Phytologist 172: 592-597; I Armstead I, Donnison I, Aubry S, Harper J, Hörtensteiner S, James C, Mani J, Moffet M, Ougham H, Roberts L, Thomas A, Weeden N, Thomas H, King I. 2007. Cross-species identification of Mendel's I locus. Science 315: 73.

[217] See note 214.

[218] Barbara McClintock (1902–1992), discoverer of transposable elements and one of only two or three plant scientists ever to have won a Nobel Prize. Among many astonishing facts about her illustrious life is the story that, while a student, she played banjo in a jazz band. Banjo, well, I don't know about that. But jazz band, now that's something I certainly approve of.

[219] See Distelfeld A, Avni R, Fischer AM. 2014. Senescence, nutrient remobilization, and yield in wheat and barley. Journal of Experimental Botany 65: 3783-3798.

[220] Harold William Woolhouse (1932-1996) was one of the Big Beasts of plant senescence research, in more ways than one. He was a tall, imposing figure and adopted a donnish,

faintly menacing demeanour that concealed an inner subversive. He once told me of his trek in the Sierra Nevada to visit the grove of ancient bristlecones (see notes 130, 133). He carried with him a small plastic feeding chamber and a vial containing a solution of radioactive bicarbonate, with the object of satisfying himself that the long-lived leaves of these venerable trees were capable of photosynthesis. He explained that, unsurprisingly, they were but the thing about the experiment that gave him the greatest satisfaction was anticipating what it would do to future attempts at carbon dating the subject of his study. The quotation is from Woolhouse's paper given at a Society for Experimental Biology Symposium on Ageing, the published proceedings of which (edited by Woolhouse) in some ways marked the beginning of the modern era of senescence research: Woolhouse HW (editor). 1967. *SEB Symposium 21: Aspects of the Biology of Ageing* (Cambridge University Press).

[221] Systems Biology is the discipline that seeks to analyse the regulatory networks that modulate gene expression in response to developmental and environmental signals. For an account of the application of Systems Biology tools to senescence, see Penfold CA, Buchanan-Wollaston V. 2014. Modelling transcriptional networks in leaf senescence. Journal of Experimental Botany 65: 3859-3873.

[222] Thomas Andrew Lehrer (born 1928), the great American satirist, composer and performer. Also mathematician (he published two papers in the 1950s).

[223] For a discussion of juvenility and maturity in relation to competence to senesce, see Thomas (2013 - note 4).

[224] See Robertson AL, Wolf DE. 2012. The role of epigenetics in plant adaptation. Trends in Evolutionary Biology, 4(1), e4 (doi:http://dx.doi.org/10.4081/eb.2012.e4).

[225] Rapamycin takes its name from Rapa Nui, the Polynesian inhabitants of Easter Island. Suren Sehgal discovered that *Streptomyces hygroscopicus*, a bacterium isolated from soil beneath a massive Easter Island stone head (moai), produced a powerful anti-fungal compound (Vézina C, Kudelski A, Sehgal SN. 1975. Rapamycin (AY-22,989), a new antifungal antibiotic. Journal of Antibiotics 28: 721–726). There's a story that, while moonlighting on the properties of this potentially useful antibiotic, Sehgal used it to create an ointment that cured a neighbour's fungal skin infection. Subsequently rapamycin became well established as an immunosuppressant. Its anti-ageing properties, however, are more controversial (Neff F, Flores-Dominguez D, Ryan DP, Horsch M. and many others. 2013. Rapamycin extends murine lifespan but has limited effects on aging. Journal of Clinical Investigation 123: 3272-3291).

[226] See note 124

[227] Sir William Schwenck Gilbert (1836-1911), playwright, librettist, poet. Most of the great lyricists of the golden age of American theatre and popular songs acknowledge the

inspiration of the Savoy Operas, though Stephen Sondheim felt obliged to take a spade to the soufflé (Sondheim S. 2010. *Finishing the Hat*, Knopf). But then he doesn't like Noel Coward either. Maybe it's a case of the New Yorker's 'not invented here' sniffiness. The quotation is from 'A discontented sugar broker' in Gilbert WS. 1908. *The Bab Ballads* (Harvard University Press, 1980 edition).

[228] See Smeekens S, Hellmann HA. 2014. Sugar sensing and signaling in plants. Frontiers in Plant Science. 5: 113 (doi:10.3389/fpls.2014.00113).

[229] For example Jin Y, Ni DA, Ruan Y-L. 2009. Posttranslational elevation of cell wall invertase activity by silencing its inhibitor in tomato delays leaf senescence and increases seed weight and fruit hexose level. Plant Cell 21: 2072–89.

[230] Chapter 10 in Jones et al. 2013. *The Molecular Life of Plants* (see note 9) is a survey of plant hormones and their physiological effects. For a review of hormones and senescence, see Jibran R, Hunter DA, Dijkwel PP. 2013. Hormonal regulation of leaf senescence through integration of developmental and stress signals. Plant Molecular Biology 82: 547-561.

[231] Since this beautiful experiment was reported, it has been replicated many times in many different species. We have done it ourselves. Our variations on the Gan and Amasino theme concerned manipulating a real crop species – maize – and the use of a promoter from maize rather than from Arabidopsis: Robson PRH, Donnison IS, Wang K, Frame B, Pegg SE, Thomas A, Thomas H. 2004. Leaf senescence is delayed in maize expressing the *Agrobacterium IPT* gene under the control of a novel maize senescence-enhanced promoter. Plant Biotechnology Journal 2: 101-112. The same approach and genetic constructs also worked in ryegrass: Li Q, Robson PRH, Bettany AJE, Carver TLW, Donnison IS, Thomas H, Scott IM. 2004. Modification of senescence in ryegrass transformed with IPT under the control of a monocot senescence-enhanced promoter. Plant Cell Reports 22: 816-821.

[232] Blagosklonny MV, Hall MN. 2009. Growth and aging: a common molecular mechanism. Aging 1: 357-362.

[233] TOR is a kinase, that is, an enzyme that activates or suppresses the functions of other proteins by attaching phosphate groups to them. At the core of many of the regulatory and signalling networks in cells are cascades of kinases, each phosphorylating, and being phosphorylated by, kinases. The status of the TOR system in plants is reviewed by Xiong Y, Sheen J. 2014. The role of Target of Rapamycin signaling networks in plant growth and metabolism. Plant Physiology 164: 499-512.

[234] Clive Maine McCay (1898–1967), American nutritionist and gerontologist. He and his wife, Jeanette Beyer McCay (1902-1999) played leading roles in the New York State Emergency Food Commission, promoting the nutritional benefits of soybeans, notably in the form of 'Cornell bread' (made with soy flour), during World War II and the following decades.

[235] See, for example, Sohal RS, Forster MJ. 2014. Caloric restriction and the aging process: a critique. Free Radical Biology and Medicine 73: 366-382.

[236] It seems clear that, in mammalian and other non-plant systems, cellular symptoms of ageing are associated with, and can be induced by, diminished autophagic potential. In this context, autophagy is considered to be a cytoprotective measure, defending against cell, tissue and organ damage, thereby contributing positively to longevity - Rubinsztein DC, Mariño G, Kroemer G. 2011. Autophagy and aging. Cell 146: 682–695. But autophagy is itself also a component process of senescence physiology in many organisms, degrading and recycling the fabric of cells in a terminal state – see Avila-Ospina et al. (2014), note 207. Reconciling these contrasting functions of autophagy in development, ageing and senescence is difficult – at least, I find it so.

[237] This is the essence of Kirkwood's (1977) disposable soma hypothesis – note 30. See also Kirkwood TBL. 2002. Evolution of ageing. Mechanisms of Ageing and Development 123: 737-745.

[238] This point of view is argued in a deliberately provocative paper: Thomas H, Sadras VO. 2001. The capture and gratuitous disposal of resources by plants. Functional Ecology 15: 3-12. I don't think Victor, my co-author, was ever fully comfortable with the mischievous intention of the piece. It was fun to write and defend it, though, and I stand by the broad conclusion – that plants are generally resource- and energy-rich and organise their lives accordingly.

[239] 'Judge a tree from its fruit, not from its leaves' (attributed to the Greek playwright Euripides, c 480-406 BCE). But the sequence of colour changes during ripening of tomatoes or bell peppers gaudily epitomises the transitions that leaves and other green organs (including fruits) undergo during senescence.

[240] The quotation is from Kamkwamba W, with Mealer B. 2009. *The Boy Who Harnessed the Wind* (HarperCollins). William Kamkwamba (born 1987) is an inspirational inventor and author from Malawi who has found fame through the innovative engineering achievements described in his book, and through many features across the range of broadcast and print media, including appearances at TED (Technology, Entertainment, Design) events. *William and the Windmill*, a film of Kamkwamba's journey from his home village in Malawi to graduation from Dartmouth College in the USA, won the Grand Jury Award for Documentary Feature at SXSW 2013 (the South By Southwest Film Festival) held in Austin, Texas.

[241] According to Danny Dorling, Thomas Robert Malthus (1766-1834) was the world's first salaried economist (Dorling D. 2013. *Population 10 Billion*, Constable). Malthus's book *An Essay on the Principle of Population*, published in 1798, influenced the development of Alfred Russel Wallace and Charles Darwin's theory of natural selection.

[242] These points are developed at length in Thomas H, Ougham H. 2015. Senescence and crop performance. In: *Crop Physiology. Applications for Genetic Improvement and Agronomy.* 2nd edition (eds VO Sadras, DF Calderini) pp. 223-249 (Academic).

[243] Lloyd Thomas Evans (1927–2015), New Zealand born plant scientist of Welsh descent, is something of a hero of mine. During 50 years at the Commonwealth Scientific and Industrial Research Organisation, (CSIRO) in Canberra, he carried out important work on the physiology of crops and, of particular significance for my own research, he discovered the Ceres line of the grass *Lolium temulentum* which has become a unique model for genetic and physiological studies – see Thomas et al. (2011), note 120. In January 2015, the year of Evans's death, the Lloyd T. Evans Plant Growth Facility was named in his honour at the International Rice Research Institute's (IRRI) centre in Los Baños, the Philippines. The quotation is from Evans's landmark study of the nature and origins of agricultural plants: Evans LT. 1993. *Crop Evolution, Adaptation and Yield* (Cambridge University Press). See also note 49.

[244] I first started to develop these ideas about the relationship between canopy longevity and yield in a chapter for a book on crop photosynthesis: Thomas H.1992. Canopy survival. In: *Crop Photosynthesis: Spatial and Temporal Determinants* (eds NR Baker, H Thomas) pp.2-41 (Elsevier). Functional stay-greens (Thomas and Ougham, 2014 – see note 63) have extended canopy duration and in many, but not all, cases enhanced yields. For a recent review of the agricultural implications of increasing foliar longevity, see Gregersen PL, Culetic A, Boschian L, Krupinska K. 2013. Plant senescence and crop productivity. Plant Molecular Biology 82: 603-622.

[245] Discussed in Guo Y. 2013. Towards systems biological understanding of leaf senescence. Plant Molecular Biology 82: 519–528.

[246] The terminology for NAC transcription factors active in senescence can be confusing. AtNAP is a member of the NAC family with a regulatory function in Arabidopsis leaf senescence (Guo Y, Gan S. 2006. AtNAP, a NAC family transcription factor, has an important role in leaf senescence. Plant Journal 46: 601–612). In wheat, *GPC* and *NAM* are genetic loci of the *NAC* type – see note 247.

[247] This important study on the evolution and function of *NAC* genes in wheat was carried out by Uauy C, Distelfeld A, Fahima T, Blechl A, Dubcovsky J. 2006. A *NAC* gene regulating senescence improves grain protein, zinc, and iron content in wheat. Science 314: 1298-1301. GPC (Grain Protein Content) was long known to be a heritable trait in domesticated wheats. By using molecular mapping techniques, these researchers found that the *GPC* locus coincides with the *NAC* family gene *NAM-B1* in the wheat genome. They went on to show that the variants of *NAM-B1* genes in bread and durum wheats are non-functional. Domestication of cereals at the dawn of agriculture fixed these inactive variant DNA sequences in the genetic backgrounds of wheat crop species. Because of this

inadvertent manipulation of *NAM* expression, leaf senescence in cultivated wheats is greatly delayed, resulting in deficiencies in grain protein and mineral nutrients compared with wild relatives.

[248] Colin Malcolm Donald (1910-1985), agronomist and plant breeder who introduced the concept of the ideotype in an influential paper describing the optimal wheat plant: Donald CM. 1968. The breeding of crop ideotypes. Euphytica 17: 385-403.

[249] The question as to how senescence should be integrated into crop plant ideotypes is addressed by Gregersen PL, Holm PB, Krupinska K. 2008. Leaf senescence and nutrient remobilisation in barley and wheat. Plant Biology 10 (Suppl.): 37–49; and Wu X, Kuai B, Jia J, Jing H. 2012. Regulation of leaf senescence and crop genetic improvement. Journal of Integrative Plant Biology 54: 936–952.

[250] See Lee EA, Tollenaar M. 2007. Physiological basis of successful breeding strategies for maize grain yield. Crop Science 47: S202–S215; Wu X. 2009. Prospects of developing hybrid rice with super high yield. Agronomy Journal 101: 688–695; Fischer RA, Edmeades GO. 2010. Breeding and cereal yield progress. Crop Science 50: S85-S98.

[251] Brigitte Anne-Marie Bardot, born 1934, former film star much admired by Messieurs d'un Certain Âge. Now animal activist and holder of erratic, largely right-wing, political opinions.

[252] See Li L, Yuan H. 2013. Chromoplast biogenesis and carotenoid accumulation. Archives of Biochemistry and Biophysics, 539: 102-109.

[253] Butelli E, Titta L, Giorgio M, Mock HP, Matros A, Peterek S, Schijlen EGWM, Hall RD, Bovy AG, Luo J, Martin C. 2008. Enrichment of tomato fruit with health-promoting anthocyanins by expression of select transcription factors. Nature Biotechnology 26: 1301-1308. This study is a tour-de-force of genetic engineering, but one is bound to wonder about the health and diet objectives. A glass or two of red wine would seem to be a simple traditional, and perhaps more pleasurable, way to get to the same outcome.

[254] Seymour G, Tucker GA, Poole M, Giovannoni J (editors). 2013. *The Molecular Biology and Biochemistry of Fruit Ripening* (Wiley).

[255] The quotation is, of course, from *The Importance of Being Earnest* (1895) by Oscar Fingal O'Flahertie Wills Wilde (1854–1900).

[256] See Plaxton WC, Podestá FE. 2006. The functional organization and control of plant respiration. Critical Reviews in Plant Sciences 25: 159–198; Paul V, Pandey R, Srivastava GC. 2012. The fading distinctions between classical patterns of ripening in climacteric and non-climacteric fruit and the ubiquity of ethylene—An overview. Journal of Food Science and Technology 49: 1-21.

257 Graham LE, Schippers JH, Dijkwel PP, Wagstaff C. 2012. Ethylene and senescence processes. Annual Plant Reviews 44: 305-341.

258 Many of the numbers given are from WRI (www.wri.org/ [accessed 23 February 2016]), one of the useful online gateways to information on the food system and other global resources.

259 Words and music © Howard Sidney Thomas 2015.

260 'He stubbed out one cigarette and lit another. "One has to look for different ways. One has to look for scaling structures – how do big details relate to little details...The only things that can ever be universal, in a sense, are scaling things"' – mathematician Mitchell Feigenbaum (born 1944), quoted in Gleick J. 1987. *Chaos. Making a New Science* (Heinemann).

261 Oliver Wendell Holmes (1809–1894), author and medical reformer, joked that the poem showed what happens when logic is carried to 'logical consequences' (Holmes OW. 1858. *Autocrat of the Breakfast Table*. Walter Scott).

262 I have been fascinated by Paul Gustave Louis Christophe Doré (1832-1883) from an early age, probably because of familiarity with my grandmother's Doré Bible. He was an odd man, straight out of the Freudian textbooks. For the whole of his 50 years he was his mother's room-mate and life companion and lost the will to live immediately after his mother's death. It is estimated he must have averaged six illustrations a day throughout his life. He became immensely wealthy and famous, but never produced one work in colour and was said to be incapable of drawing from nature. The illustration shows Acheron, the mythical river of woe across which the newly dead are ferried to Hell by Charon. There is a real River Acheron, arising near the Greek village of Zotiko. I find it amusing that it gets a four and a half star approval rating on the travel review website TripAdvisor.

263 Ludwig Josef Johann Wittgenstein (1889–1951), philosopher. During his time at Cambridge he liked to take a bag of pies to the cinema and eat them while sitting in the front row, because the noise and flashing lights (he particularly liked Westerns) helped him to think; Sydney Smith (1771–1845), cleric, wit and eccentric. He would ride about his parish at Foston-le-Clay in Yorkshire in 'The Immortal', an ancient green chariot drawn by a carthorse, and his butler was a girl 'made like a milestone', whom he named 'Bunch'; Marcus Aurelius Antoninus Augustus (121–180), soldier, Roman Emperor and author of the stoic text *Meditations*. He has been revered by the Great and the Good throughout history (Bill Clinton read him avidly) as a kind of Zen teacher (though his persecution of Christians and suppression of the Germanic tribes was anything but Zen-like). In his biography of Marcus Aurelius, Frank McLynn concludes that he was a decent, thoughtful man 'caught up in the whirlwind of history' – the Jan Christian Smuts of his generation (McLynn F. 2009. *Marcus Aurelius: Warrior, Philosopher, Emperor*. Bodley Head).

[264] From the 1814 blank verse translation of the *Commedia* by The Reverend Henry Francis Cary.

[265] See note 262.

[266] Acherontic changes are exploited in food, feed and fibre production processes such as retting flax for linen, malting cereal grains for brewing and ensiling forages for animal consumption.

[267] These conclusions are based on arguments developed in Thomas H. 2003. Do green plants age and if so how? Topics in Current Genetics 3: 145-171.

[268] Our death is not an end if we can live on in our children and the younger generation. For they are us; our bodies are only wilted leaves on the tree of life (Albert Einstein).

Find and defined

Y

Z

About this book

I wanted to write this book because, after two thirds of a lifetime thinking about senescence, my bonnet is full of bees, my belfry is infested with bats, my cupboard is crammed with skeletons and it's time they were purged. There's another reason. In an era when to be a jobbing scientist is to find oneself adrift in an almost boundless ocean of data and information, it's essential to have a life raft – a personal narrative with which to navigate the tides, currents and storms of this restless sea. This book is my senescence narrative. It's partial and prejudiced, but it's what keeps me from going down for the third time. And one more reason: this is the closest I intend to come to writing a memoir.

Frequently in these pages I refer to friends and colleagues. It's my way of acknowledging their influence on my researches into, and ideas about, senescence in all its diversity. I owe a further debt of gratitude to many others, fleetingly listed among my co-authors on the publications I've cited. And there are family members, friends and colleagues too, who remain unnamed but are no less deserving of thanks. You know who you are.

I wouldn't have had the foggiest about how to turn this brain-dump into a book if it weren't for Jayne Archer and Richard Marggraf Turley, who can take much credit for any scholarship (if that isn't too grand a word) to be found in these pages. I greatly appreciate support of all kinds from Aberystwyth University and the New Phytologist Trust.

Flann O'Brien didn't believe a book should have only one ending. I rarely disagree with him, but in this case, one is enough.

Howard Thomas
Aberystwyth and Wye
May 2016

About the author

Howard Thomas was born and educated in Wales and, after a career in scientific research, he is now emeritus Professor of Biology at Aberystwyth University. This allows him to carry on with all the nice parts of the old job but without any of the garbage (admin, budgets, staff and student issues, pointless meetings). It means he has been able lately to indulge a special interest in the science-humanities connection. He also leads a parallel life as a devout jazz musician.

Other books authored or edited by Howard (Sidney) Thomas

H Thomas, D Grierson, eds. 1987. Developmental Mutants in Higher Plants. Cambridge: University Press

N Baker, H Thomas, eds. 1992. Crop Photosynthesis: Spatial and Temporal Determinants. Amsterdam: Elsevier

RL Jones, H Ougham, H Thomas, SD Waaland. 2013. The Molecular Life of Plants. NJ, Chichester: Wiley

J Archer, R Marggraf Turley, H Thomas. 2014. Food and the Literary Imagination. London: Palgrave

S Thomas. 2016. 20 Steps to Jazz Keyboard Harmony. e-book published at http://tinyurl.com/hw6seg4

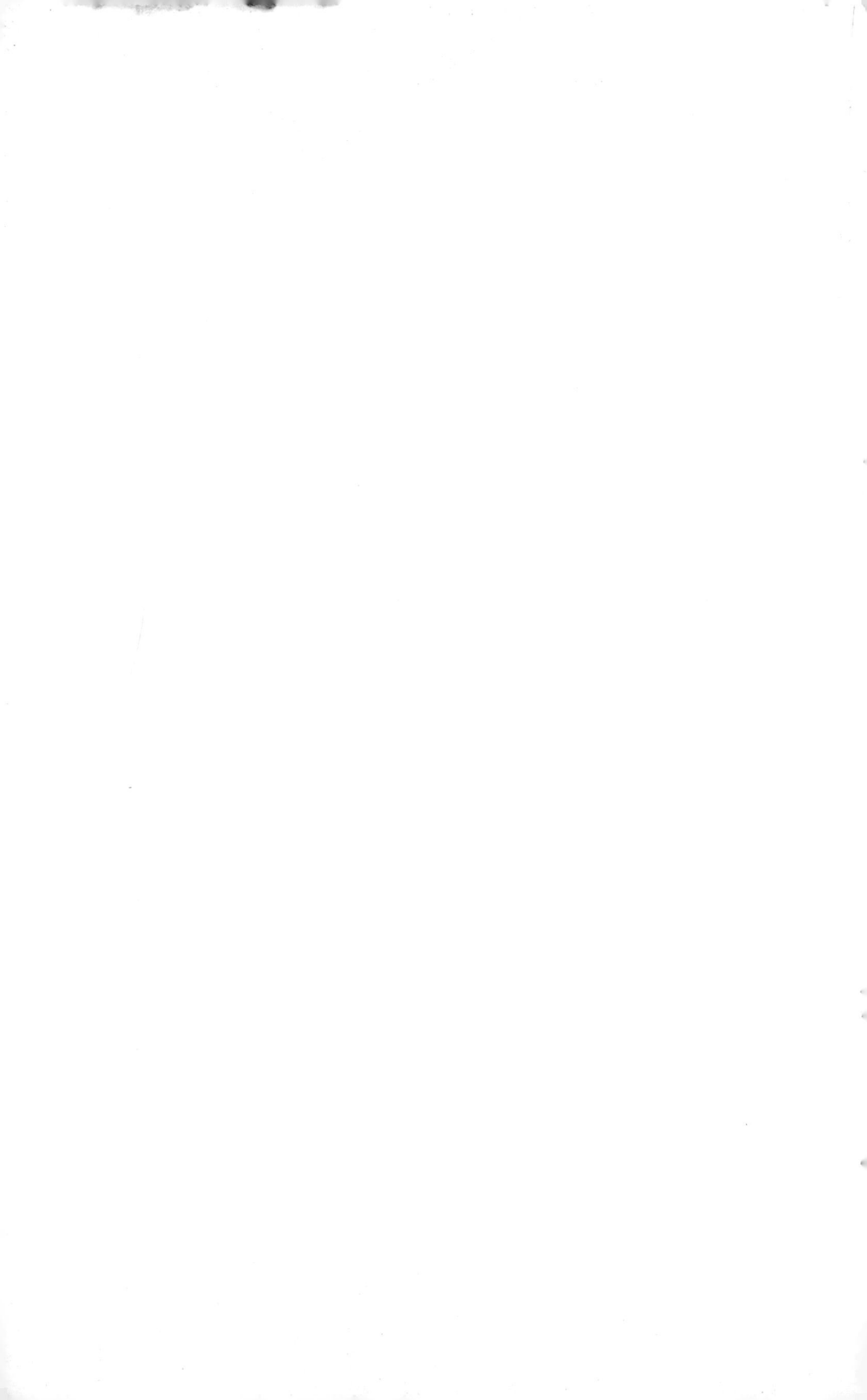